Cope's Early Diagnosis
of the Acute Abdomen

Sir Zachary Cope
1881–1974

Cope's Early Diagnosis of the Acute Abdomen

EIGHTEENTH EDITION

Revised by
WILLIAM SILEN, M.D.

Johnson & Johnson Professor of Surgery
Harvard Medical School

New York   Oxford
OXFORD UNIVERSITY PRESS
1991

Oxford University Press

Oxford   New York   Toronto
Delhi   Bombay   Calcutta   Madras   Karachi
Petaling Jaya   Singapore   Hong Kong   Tokyo
Nairobi   Dar es Salaam   Cape Town
Melbourne   Auckland

and associated companies in
Berlin   Ibadan

Published by Oxford University Press, Inc.,
200 Madison Avenue, New York, New York 10016

Library of Congress Cataloging-in-Publication Data
Silen, William, 1927–
Cope's early diagnosis of the acute abdomen. — 18th ed.
/ revised by William Silen.
p.   cm.   Includes index.
ISBN 0-19-506734-7 (cloth). — ISBN 0-19-506735-5 (pbk.)
1. Acute abdomen—Diagnosis. I. Cope, Zachary, Sir, 1881–1974
Early diagnosis of the acute abdomen. II. Title.
III. Title: Early diagnosis of the acute abdomen.
[DNLM: 1. Abdomen, Acute—diagnosis.   WI 900 S582c]
RD540.S495   1991   617.5′5075—dc20   DNLM/DLC   90-7759

9 8 7 6 5 4 3 2 1

Printed in the United States of America
on acid-free paper

lenge of revising this book. I am particularly grateful to Ms. Nancy Kaufman and to Ms. Pamela Dale for their superb assistance in preparing the manuscript.

Finally, I should like to pay special tribute to all who have taught me about acute abdominal pain, including my teachers, my pupils, and particularly my patients.

Boston   W. S.
July 1990

Contents

Plates

1. Parietal muscles of the abdomen

RADIOGRAPHS

2. Free gas between the diaphragm and the liver in a case of perforated peptic ulcer

3. Perforated appendicitis with peritonitis in a child with a classical fecolith

4. Ultrasound examination of the gallbladder, showing calculi and "shadows" beyond the stones

5. Ultrasound examination showing a very large abdominal aneurysm in transverse section

6. Ileal obstruction by adhesions

7. Multiple fluid levels in the small intestine in a case of intestinal obstruction

8. Series of radiographs taken during the administration of a barium enema in a case of intussusception

9. Volvulus of the cecum

10. Volvulus of the sigmoid

11. Barium enema of the patient shown in Plate 10

12. Plain film showing free air in the biliary tree in a patient with gallstone obstruction of ileum

13. Obstruction of the colon by a carcinoma of the sigmoid

14. Barium enema showing a pericolic abscess and a fistulous tract arising from sigmoid diverticulitis

There is surely no greater wisdom than well
to time the beginning and onsets of things.

BACON, "On Delay"

Cope's Early Diagnosis
of the Acute Abdomen

1. The principles of diagnosis in acute abdominal disease

Before entering on the detailed consideration of the various forms of acute abdominal pain, it is well to lay down certain principles that form the basis of all successful diagnosis in urgent abdominal disease.

Necessity of making a diagnosis

The first principle is that of the *necessity of making a serious and thorough attempt at diagnosis, usually predominantly by means of the history and physical examination.*

Abdominal pain is one of the most common conditions that calls for prompt diagnosis and treatment. Usually, though by no means always, other symptoms accompany the pain, but in most cases of acute abdominal disease pain is the main symptom and complaint. The very terms "acute abdomen" and "abdominal emergency," which are constantly applied to such cases, signify the need for prompt diagnosis and early treatment, by no means always surgical. *The term "acute abdomen" should never be equated with the invariable need for operation.* It is common knowledge, however, that when confronted with a patient suffering great abdominal pain, it is often difficult to be certain about the exact intraabdominal lesion that has given rise to the symptoms. In some instances the urgent need for operation may be so obvious that the need of transference of the patient to the care of a surgeon is clear. In other cases the observer may, if in doubt, think it wise to discuss the problem with a fellow practitioner before deciding on any course of action. There are, however, occasions when, with somewhat indefinite symptoms, there is justification to wait for the development of clearer indications, to see if the condition will not improve spontaneously, and to temporize, so long as the patient is carefully observed at frequent intervals. Though in some cases it is impossible to be certain of the diagnosis, it is a good habit to come to a decision in each case; and it will be found that after a short time, the percentage of correct diagnoses will rapidly increase.

That there is much room for improvement in this direction cannot be gainsaid. The operating surgeon is not free from blame in this matter, for the ease and comparative safety of operating, and the ready availability of computerized tomography (CT) and ultrasound examinations, occasionally cause him to make a rather perfunctory examination of some patients whom from previous experience he judges to be in urgent need of laparotomy. If every physician were to make a thorough attempt at a full diagnosis before operating, the science of elucidation of acute abdominal disease would be advanced considerably. There is no field in which diagnosis should be so precise, since in no class of cases has the physician so great an opportunity of correlating the symptoms with the pathology.

It is only by thorough history taking and physical examination that one can propound a diagnosis, and if the early stages of the disease are to be recognized, note must be taken of the earliest symptoms. General practitioners have better opportunities than any other section of the medical community for observing these early symptoms, and by patient and painstaking observation it is possible for them to add greatly to the stock of common knowledge. To attempt a specific diagnosis prevents carelessness, and carelessness in urgent abdominal diagnosis is close akin to callousness.

It is a truism to say that correct diagnosis is the essential preliminary to correct treatment. Many serious results have followed from an observer's jumping to wrong conclusions that might easily have been avoided by a real attempt at clinical differentiation.

Spot diagnosis may be magnificent, but it is not sound diagnosis. It is impressive but unsafe. The deduction and induction from observed facts necessary for the formation of a definite opinion provide good mental discipline for the observer, help to imprint upon the tables of the mind perceptions and clinical pictures that can usefully be recalled in the future, and give a sense of satisfaction that is only slightly diminished if the resulting opinion should prove to be incorrect. One often, if not always, learns more by analyzing the process of and detecting the fallacy in an incorrect diagnosis than by taking unction to oneself when the diagnosis proves correct.

Early diagnosis

There can be no question that in acute abdominal disease it is of the utmost importance to *diagnose early*. Like the businessman who takes as his motto "Do it now," the medical man, when confronted with an urgent abdominal case, should have ever before him the words "Diagnose now." The patient cries out for relief, the relatives are insistent that something shall be done, and the humane disciple of Aesculapius is driven to diminish or banish the too-obvious agony by administering a narcotic. There is little doubt that the giving of narcotics by a physician *not* responsible for the ultimate diagnosis and therapy may lead to serious delay and error in decision making. The realization that this can obscure the clinical picture has given rise to the unfortunate dictum that narcotics should never be given until a diagnosis has been firmly established. With the numerous layers of triage nurses, medical students, residents, and attending physicians in modern emergency units, and with the addition of time-consuming tests often done before an adequate history *and* physical examination, the suffering patient is sometimes forced to wait for many hours before any relief is offered. This cruel practice is to be condemned, but I suspect that it will take many generations to eliminate it because the rule has become so firmly ingrained in the minds of physicians.

The ideal solution to this problem is for a *responsible surgeon* to evaluate the patient at the earliest possible time. An adequate history, pertinent physical examination, and a tentative working diagnosis can be accomplished by an experienced surgeon in a relatively short period of time, after which relief should be given without hesitation. Should any tests be required, these can then be done with greater comfort for the patient. It is the examination, reexamination, and testing ordered by individuals inexperienced in the diagnosis of abdominal pain that leads to delay in diagnosis and failure to provide early relief of pain.

It is a curious but well-known fact that many who develop abdominal pain in the daytime endure until evening before they feel compelled to come to the hospital. It follows that important decisions often have to be made at night when the physician, weary with the day's work, and with perceptions and reasoning faculties somewhat jaded, is both physically and mentally below his best. The temptation is often very strong to temporize and "see how things are in the morning." There can be few practitioners of experience who cannot look back with regret to one or more occasions when delay has been fraught with disaster. The waiting attitude is understandable, but only occasionally excusable. To suspect an intussusception, to think that possibly there may be a perforation of a gastric ulcer, and yet to leave the question undecided for eight or ten hours is to gamble with a life. The fact that the patient comes late to see the doctor is all the more reason why he should diagnose as soon as possible. *The general rule can be laid down that the majority of severe abdominal pains that ensue in patients who have been previously fairly well, and that last as long as six hours, are caused by conditions of surgical import.* There are exceptions, but the generalization is useful if it serves to call attention to the need for early diagnosis. Surgical intervention is usually required for perforated ulcers, for appendicitis, for intestinal obstruction, and for many other conditions, and it is well recognized that the earlier such conditions are dealt with by the surgeon, the better the results. But the old view that delay is permissible still lingers in some quarters, for custom changes slowly. Public opinion on the subject needs education, and such education must come from the practitioner. Since this book was first published, however, the benefit of surgical intervention in acute abdominal conditions has become recognized by the public, and it is rare to find patients who refuse to undergo any operation that is strongly advised by the surgeon.

The recovery rate from acute abdominal disease increases in proportion to the earliness of diagnosis and treatment. During the past several decades there has been a considerable reduction in the mortality of acute abdominal diseases, especially acute appendicitis and other infective processes such as perforated diverticulitis. This may have been due to several causes—the introduction of antibiotics or the increase in the number of trained surgeons—but we believe it likely that some of the improvement may have been due to earlier diagnosis.

There has also been an improvement, though not to the same extent, in the results of operative treatment of intestinal obstruction and strangulated hernia. There is, however, still considerable room for improvement in the diagnosis of the early symptoms of intestinal obstruction and in the realization that operation is safer than prolonged attempts to reduce a strangulated hernia manually.

Thorough history and physical examination

The necessity of making a *thorough physical examination* in every acute abdominal case should not need much emphasis. Radiologic or ultrasonic examinations, CT, and the vast array of laboratory tests available to all of us today will not compensate for a poor or incomplete history and physical examination. If one is to make a correct diagnosis, a complete history and physical examination should be the rule. It is as important to insert the finger into the lower end as to order a plain film of the abdomen. More early cases will be diagnosed by palpating the pelvic peritoneum than by perusing the CT scan. In the most perfunctory examination one is almost bound to lay the hand on the patient's abdomen, and if the latter is tender and rigid, the assumption may be made that the condition is a local peritonitis, though a stethoscope applied to the lower part of the chest might possibly reveal the fact that the origin of the symptoms was a diaphragmatic pleurisy.

The exact order or method of examination which one may follow is a matter of individual choice or preference, but the routine followed by the writer is indicated and described in the succeeding chapters.

Knowledge of anatomy

Many examinations of the abdomen are imperfect because the practitioner does not act upon the important *principle of applying his knowledge of anatomy.* It is well to cultivate the habit of thinking anatomically in every case where the knowledge of structural relations can be put to advantage. There are very few abdominal cases in which this cannot be done. Application of anatomy makes diagnosis more interesting and more rational. The explanation of some doubtful point, the differentiation of the possible causes of a pain, and the determination of the exact site of a diseased focus often depend upon small anatomical points.

One can best illustrate the value of applied anatomy in abdominal diagnosis by considering those structures that are least variable in their position—the voluntary muscles and the cerebrospinal nerves. Plate 1 shows well the position of the different muscles, the diaphragm, the psoas, the quadratus lumborum, the erector spinae, the pyriformis, and the obturator internus. Each of these muscles may be of great clinical significance, for if any of them is irritated directly or reflexly by inflammatory changes it becomes tender and rigid, and pain is caused when the muscle fibers are moved. Everyone is acquainted with the rigidity of the rectus and lateral abdominal muscles when there is a subjacent inflammatory focus, but few take much note of the rigidity of the diaphragm in a case of subphrenic abscess because the diaphragm is invisible and impalpable. Its immobility may be deduced, however, from the impairment of movement of the upper part of the abdominal wall and the lower thorax.

It will be remembered that in some cases of appendicitis and other conditions affecting the psoas muscle there is flexion of the thigh, due to contraction of the muscle consequent on direct or reflex irritation, but how often does anyone test the slighter degrees of such irritation by causing the patient to lie on the opposite side and extending to the full the thigh on the affected side? Again, the obturator internus is covered by a dense fascia and is not readily irritated by pelvic inflammation; but if there is an abscess (e.g., one caused by a ruptured appendix) immediately adjacent to the fascia, pain will be caused if the muscle is put through its full movement by rotating the flexed thigh inward to its extreme limit. The pain is referred to the hypogastrium. This sign is not present in every case of pelvic appendicitis and may occur in other pelvic conditions, such as pelvic hematocele, but when present it denotes a definite pathological change (Fig. 1).

The application of the knowledge of the anatomical course and distribution of the segmental nerves is also important. When a patient complains of loin pain radiating to the corresponding testis, one remembers the embryological fact that the testis is developed in the same region as the kidney; and though the former travels to the scrotum just before birth, in suffering it shows its sympathy with and serves as an indicator for the intra-abdominal structure which was developed near it. Of course, pain referred to the testicles does not always denote primary genitourinary disease. It is probable that the main nerve supply to the vermiform appendix comes from the tenth dorsal segment, so that pain in one or both testicles may be caused by such a condition as appendicitis. The dorsal distribution of referred pain should also be noted (Fig. 2).

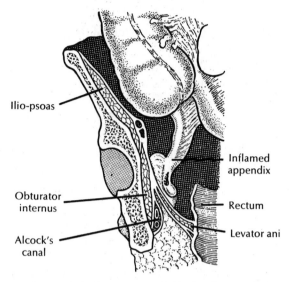

Fig. 1. Anatomical parts concerned in the obturator test.

Another segmental pain of great importance is that referred from the diaphragm. The diaphragm begins to develop in the region of the fourth cervical segment, from which is obtained the major part of its muscle fibers. Nerve fibers, mainly from the fourth cervical nerve, accompany the muscle fibers and constitute the phrenic nerve. The growth of the thoracic contents causes the muscle to be displaced caudalward, and it finally takes up its position at the lower outlet of the thorax. The phrenic nerves elongate to accommodate themselves to the displaced muscle. From the diagnostic point of view the separation from the original position is extremely valuable, for if in upper abdominal or lower thoracic lesions pain is felt, or hyperesthesia is detected in the region of distribution of the fourth cervical nerve, it is a strong presumption that the diaphragm is irritated by some inflammatory or other lesion. The significance in abdominal diagnosis of constant or intermittent pain in the region of distribution of the fourth cervical nerve is still sometimes either not understood or seriously neglected. Pain on top of the shoulder may be the only signal that an inarticulate liver abscess,

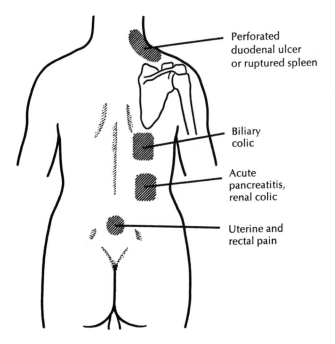

Perforated
duodenal ulcer
or ruptured spleen

Biliary
colic

Acute
pancreatitis,
renal colic

Uterine and
rectal pain

Fig. 2. The sites on the posterior surface of the body to which pain is referred in acute abdominal conditions.

threatening to perforate the diaphragm, may be able to produce. When a gastric ulcer perforates, the escaping fluid may impinge on the lower surface of the diaphragm, irritate the terminations of the phrenic nerve on one or both sides, and cause pain on top of one or both shoulders. Pain may also be referred to the shoulder in cases of subphrenic abscess, diaphragmatic pleurisy, acute pancreatitis, ruptured spleen, and in some cases of appendicitis with peritonitis. The pain is felt either in the supraspinous fossa, over the acromion or clavicle, or in the subclavicular fossa (Fig. 3). Pain on top of both shoulders indicates a median diaphragmatic irritation. It is not sufficient to elicit from the patient that there is pain in the shoulder, since pain at the root of the neck is sometimes taken to be pain at the tip of the scapula because of a less than meticulous history. The shoulder pain is also apt to be overlooked because the patient

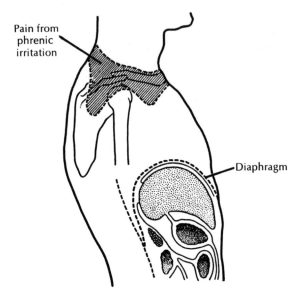

Pain from phrenic irritation

Diaphragm

Fig. 3. The shoulder area in which phrenic irritation may cause pain.

attributes it to "arthritis." Sometimes, when the diaphragm is irritated by a neighboring lesion, such as blood from a ruptured spleen, tenderness may be elicited by pressure on the corresponding phrenic nerve in the neck.

Errors in diagnosis also result from want of appreciation of another anatomical point (i.e., the lack of representation in the muscular abdominal wall of the segments that form the pelvis), so that irritation of the pelvic nerves (e.g., from pelvic peritonitis) causes no abdominal wall rigidity. Peritonitis commencing deep in the pelvis may therefore be unaccompanied by any rigidity of the hypogastric abdominal wall.

The general principle may be laid down that rigidity of the muscular wall of the abdomen in response to irritation varies in proportion to the extent of the somatic nerve supply to the subjacent peritoneum. Indeed, the whole peritoneal lining of the abdomen may be divided into *demonstrative* and *nondemonstrative* areas, the former causing reflex muscular rigidity and the latter not

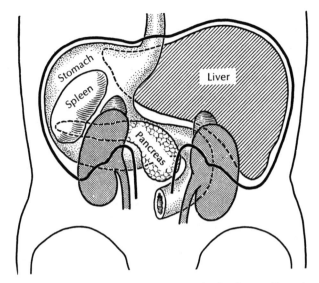

Fig. 4. The relationship of the viscera with the diaphragm. (Posterior view with the back part of the diaphragm cut away.)

doing so. The pelvis and the central part of the posterior abdominal wall may be the site of an acutely inflamed appendix or an abscess, and yet there may be no overlying rigidity of the muscles—hence the difficulty of diagnosing a pelvic or a retroileal inflamed appendix.

Figures 1–4 demonstrate the importance of applying knowledge of anatomy to abdominal diagnosis. It is unnecessary to emphasize the great importance that a thorough acquaintance with the normal size, position, and relations of the abdominal viscera has in connection with the elucidation of abdominal disease.

Knowledge of physiology

The localization of inflammatory lesions is aided particularly by knowledge of anatomy, while in obstructive lesions the application of physiological knowledge plays perhaps a more important part.

A very large number of urgent abdominal cases are accompanied by pain due to abnormal conditions in tubes whose walls are composed mainly of smooth muscle fibers. There is no high-grade sensibility in such tubes and inflamed abdominal viscera are not necessarily tender on palpation. It is possible to crush, cut, or tear intestine without the fully conscious patients experiencing any pain, yet everyone is aware that pain does take origin from the intestine. *The required stimulus is stretching or distention of the tube or excessive contraction against resistance.* Evidence of pain arising from a tube of involuntary muscle is therefore indicative of local distention, either by gas or fluid, or of vigorous contraction. In mild degrees in the intestine this is commonly called flatulence; in greater degrees in intestinal, renal, and uterine tubes it is called colic. Severe colic always indicates obstruction, sometimes only temporary, causing local distention or violent peristaltic contraction. It occurs in *paroxysms,* and the pain, which is often of an excruciating nature, is referred to the center from which the nerves come and also to the segmental distribution corresponding to the part of the spinal cord from which the sympathetic nerves to the affected viscus are derived. Colic of the small intestine causes pain referred chiefly to the epigastric and umbilical regions, while large-intestine colic is usually referred to the hypogastrium. The pain of biliary distention is usually felt more in the right subscapular region, while that of renal colic is felt in the loin and sometimes radiates to the corresponding testicle. Severe colic is one of the most terrible trials to which a human being can be subjected. Patients tend to fling themselves about, twist, and double themselves up in a characteristic way. If, therefore, a patient gets paroxysms of pain that are accompanied by the most violent restlessness of agony, the chances are that the condition is some form of obstruction and not peritonitis, for in the latter condition movement generally increases the pain.

Another physiological fact of importance is that tenderness due to irritation of nerves by a unilateral lesion is not usually felt on the opposite side of the body. For example, a right-sided pleurisy will sometimes cause tenderness and rigidity in the right but not in the left iliac region. If the fingers pressed well into the left iliac fossa and pushed deeply toward the right side across the middle line evoke tenderness, this indicates a deep inflammatory lesion in the right iliac region.

True shock as we define it today does not occur as a result of pain alone. Thus, while the patient early in the course of an acute perforated ulcer or the patient with "biliary colic" may have tachycardia and appear pale and diaphoretic, the blood pressure will usually be normal or even raised because of release of cate-cholamines. *True shock within minutes or an hour or two of the onset of abdominal pain usually means intra-abdominal hemor-rhage.* Shock that occurs in the later stages of intra-abdominal catastrophes is the result of a loss of intravascular volume from multiple causes: (1) vomiting or diarrhea, (2) sequestration of fluid into distended intestine, (3) sequestration of fluid into the inflamed peritoneal cavity, or (4) bleeding into infarcted intestine.

Visceral pain

There are certain important facts that must be constantly remem-bered in connection with sensation and pain felt in the small and large intestines and their diverticula—the vermiform appendix, Meckel's diverticulum, and the numerous diverticula so common in the large bowel.

The intestines themselves are insensitive to touch and to inflam-mation that does not affect the enclosing peritoneum. Neverthe-less, severe pain may emanate from any part of the intestine when it is severely distended or when its muscle contracts violently, but this pain is always referred somewhere along the midline of the abdomen in front except when the affected part is close to or in contact with the abdominal parietes, which is well supplied with very sensitive somatic nerves. These remarks are particularly important in connection with acute inflammation of the appendix and of the gallbladder. The interior of the gallbladder may become inflamed, irritate the mucosa, and cause the secretion of mucus, so that the viscus may be fully distended and cause central abdominal referred pain. Yet there may be no pain on manipulation of and pressure on the distended gallbladder itself; the central pain may subside, while the inflammation remains and gradually extends through the visceral wall until the products of inflammation or the bacteria themselves penetrate the wall of the gallbladder, irritate the parietal peritoneal nerves, and cause severe local pain.

Adrenal steroids and acute abdominal disease

Patients under treatment by adrenal steroids may be affected by that treatment in one or more ways that concern acute abdominal disease.

1. It is well known that adrenal steroid therapy diminishes the symptoms produced by inflammation. *Any acute inflammation may arise, with either localized or generalized peritonitis, without the usual symptoms and signs of such inflammation being sufficiently clear to cause alarm.* Even perforation of a gastric ulcer may provoke only mild discomfort in the early stages. Fever usually is completely suppressed by the steroids. The assessment of abdominal pain in persons on steroid therapy is therefore very difficult, and laparotomy may be desirable in ambiguous cases.

2. Rather inconclusive evidence has been adduced to show that patients under treatment by adrenal steroids are more prone to develop peptic ulcer or complications of an existing ulcer. The evidence is strongest in patients suffering from rheumatoid arthritis who are being treated by high doses of corticosteroids, and these patients need to be observed especially for the presence of gastric ulcer. It appears that the nonsteroidal anti-inflammatory drugs taken by these patients are more ulcerogenic than the steroids.

3. Recently, a variety of bizarre acute perforations of the small

intestine and colon has been reported in patients on long-term corticosteroid therapy who have not had previous intestinal disease. The cause of these perforations is not known, but they have occurred more often in patients with collagen diseases.

Warning concerning patients taking corticosteroids

From the above-mentioned statements it follows that the attenuation of abdominal pain in patients being treated by corticosteroids makes it imperative to consider even slight abdominal pain as serious. Neglect of this warning may have serious results, and exploration of the abdomen may be required sooner than for patients not undergoing treatment by corticosteroids.

Exclusion of medical diseases

In diagnosing actue abdominal disease it is *always necessary to exclude medical diseases* before concluding that the condition is one needing surgical intervention. Certain aspects of disease that may simulate intra-abdominal catastrophes can only be learned either in the medical wards of a large hospital or in an extensive general medical practice. Cardiac derangements, chronic interstitial nephritis and arteriosclerosis, cirrhosis of the liver, tuberculosis, peritonitis—all these and many other medical diseases will sooner or later cause doubt in abdominal diagnosis. The physician who wishes to be prepared thoroughly to examine and correctly to diagnose patients with acute abdominal pain must be thoroughly conversant with these diseases and their protean manifestations.

Opening of the abdomen is not to be advised with too light a heart. The dextrous hand must not be allowed to reach before the imperfect judgment. Laparotomy is only to be made on the recommendation of a physician of mature judgment after a thorough examination. It is regrettable to have to say afterward that one did not know that there was severe albuminuria or that the lungs were not examined.

If, however, after careful examination one comes to the conclusion that there is within the abdomen the early stage of a pathological process that tends to get worse and is amenable to surgical treatment, there should be no hesitation in recommending operation. Correct diagnosis is the basis of firm counsel.

The effect of antibiotics

It may not be out of place to sound a note of caution as to the possible effect of antibiotics on the symptoms of inflammation within the abdomen. These drugs are in such common use (sometimes when diagnosis is uncertain) that the observer, before making a diagnosis, should ask whether any of these drugs have been recently administered. If so, extra care should be taken in examination and allowance made for any possible action of the drug. Antibiotics cannot seal a perforation of the appendix, but they can diminish the symptoms of the ensuing peritonitis.

2. Method of diagnosis: the history

In diagnosing acute abdominal disease it is well to have a routine method of examination, not to be slavishly followed but to be modified according to circumstances.

History of the present condition

It can be confidently asserted that a large number of acute abdominal conditions can be diagnosed by considering carefully the history of their onset. That is only possible, however, when each symptom is carefully appraised in relation to the other symptoms, so that its significance is properly understood. *This is of the greatest use when the chronological appearance of symptoms, from the time of the last period of well-being, is meticulously recorded.* More diagnoses will be made by this means than by the CT scan.

AGE

To know a patient's age is helpful, since the incidence of certain conditions is limited to a particular range of years. Acute intussusception in temperate climates occurs generally in infants under two years of age. Obstruction of the large intestine by a cancerous stricture is seldom seen before thirty, is infrequent before forty, but is relatively common after forty. Actue pancreatitis is seldom seen in those under twenty. A perforated gastric ulcer is a rarity in persons under fifteen years. Conditions such as cholecystitis or a twisted pedicle of an ovarian cyst may occur in childhood, though much more commonly in adult life. All the acute conditions that are due to derangements of the developing ovum or its surroundings are naturally found only in women during the childbearing period.

EXACT TIME AND MODE OF ONSET

It is frequently possible for the patient to fix the exact time at which the pain started. The awakening out of sleep by acute abdominal pain is so startling that it is not forgotten, and such pain is almost always of great significance. Actue appendicitis very commonly, and perforation of a gastric or duodenal ulcer not infrequently, commence in this manner. It is no ordinary pain which begins thus.

Many acute conditions appear to be precipitated by some slight exertion or by the energetic effect of an aperient. Many cases of incipient appendicitis become much worse soon after the administration of castor oil or its equivalent. The temporary increase in intra-abdominal tension caused by any slight straining effort may cause the giving way of the thin floor of a gastric ulcer or the rupture of a pregnant fallopian tube.

It is also important to determine whether the condition began immediately after some injury; apparently trivial abdominal injuries may be accompanied or followed by serious lesions. The spleen is sometimes ruptured by comparatively slight violence applied to the lower left chest or left hypochondrium. A violent cough may fracture a costal cartilage and give rise to the upper abdominal pain. *It is useful to ask: what were you doing when the pain began?*

Acuteness of onset. The acuteness of onset gives some indication of the severity of the lesion. Ask if the patient fainted or fell down and collapsed at the onset of symptoms. Perforation of a gastric or duodenal ulcer, acute pancreatitis, and ruptured aortic aneurysm are the only abdominal conditions likely to cause a person to faint.

In a woman the rupture of an ectopic gestation also may cause fainting.

Many cases of intestinal obstruction are gradual in onset, culminating in an acute crisis. Strangulation of the gut, however, is accompanied by very acute symptoms from the first. The symptoms due to torsion of the pedicle of an ovarian cyst are also usually acute from the start.

To know the exact time of onset is useful in estimating the probable pathological changes that have ensued. For example, it is not usual for an appendix to perforate within twenty-four hours of the onset of symptoms; so, unless the symptoms definitely point to diffuse peritonitis, one may give a better prognosis for appendectomy if the operation is performed within that time. Similarly, a rigor and high fever within twenty-four hours of the onset of pain almost exclude the possibility of appendicitis. Again, it is only by ascertaining the precise moment of onset that one can tell whether the apparent well-being of the patient corresponds to the stage of reaction which is seen in some conditions—notably in patients with perforated gastric or duodenal ulcer.

PAIN: ONSET, DISTRIBUTION, AND CHARACTER

The greatest importance attaches to the very careful consideration of the onset, distribution, and character of the pain.

Situation of the pain at first. When the peritoneal cavity is flooded suddenly by either blood (from a ruptured tubal gestation sac) or pus (e.g., from a ruptured pyosalpinx) or acrid fluid (from a perforated gastric ulcer), the pain is frequently said to be felt "all over the abdomen" from the first. But the maximum intensity of pain at the onset is likely to be in the upper abdomen in the latter and in the lower abdomen in the two former conditions. In perforated duodenal ulcer the pain may be at first more acute in the right hypochondrium and right lumbar and iliac regions, owing to the irritating fluid passing down chiefly on the right side of the abdomen.

Pain arising from the small intestine, whether due to simple colic, organic obstruction, or strangulation, is always felt first and chiefly in the epigastric and umbilical areas of the abdomen (i.e., in the zone of distribution of the ninth to eleventh thoracic nerves, which supply the small intestine via the common mesentery). Remembering that the appendicular nerves are derived from the same source as those that supply the small intestine, it is not surprising that the pain at the onset of an attack of appendicitis is usually felt in the epigastrium. When small intestine is adherent to the abdominal wall, pain caused by its peristaltic movement is referred to that part of the abdominal wall to which the gut is adherent.

The pain of large-gut affections is more commonly felt at first in the hypogastrium or, in the case of the cecum and ascending colon, when the mesocecum or mesocolon is very short or wanting, at the actual site of the lesion.

The shifting or localization of the pain. This is often significant. After a blow on the upper part of the abdomen, if local pain at the site of injury is the first complaint, but in a few hours the pain is referred more to the hypogastrium, one must suspect rupture of intestine and consequent gravitation of the escaping fluid to the pelvis. Similarly, localization of pain in the right iliac fossa some hours after acute epigastric pain is usually due to appendicitis— though occasionally the same sequence is seen with a perforated pyloric or duodenal ulcer or with acute pancreatitis. When severe pain is first felt in the thorax but later is felt more in the abdominal cavity, one must at least consider the possibility of a dissecting aneurysm.

The character of the pain. This often helps to indicate the nature or seriousness of the condition. The general burning pain of a perforated gastric ulcer, the agony of an acute pancreatitis, the sharp constricting pain that takes away the breath in an attack of biliary colic, the tearing pain of a dissecting aneurysm, and the gripping pain in many cases of intestinal obstruction contrast with the acute aching in many cases of appendicitis with abscess or the constant dull fixed pain of a pyonephrosis.

Radiation of the pain. This is sometimes diagnostic. It is especially true of the colics, in which pain radiates to the area of distribution of the nerves coming from that segment of the spinal cord that supplies the affected part. Thus in biliary colic the pain is frequently referred to the region just under the inferior angle of the right scapula (eighth dorsal segment), while renal colic is frequently felt in the testicle on the same side.

In many conditions of the upper abdomen and lower thorax, pain is referred to the top of the shoulder on the same side as the lesion (see Chapter 1), and special inquiry should always be made as to pain or tenderness over the supraspinous fossa, the acromion, or the clavicle.

It is always well to *ascertain if the pain is influenced by respiration.* Pleuritic pain is usually worse on taking a deep inspiration and is diminished or stopped during a respiratory pause. Inflammation of the gallbladder may cause inhibition of movement of the diaphragm, and the pain may be increased by a forced respiration. In many cases of peritonitis, intraperitoneal abscess, or abdominal distention due to intestinal obstruction, pain will be caused or increased on inspiration.

Special varieties of pain. It is necessary to ask if there is any pain during the act of micturition, for the presence or absence of such pain is frequently of great significance. In addition to its causation by primarily urinary conditions (e.g., pyelitis), stone in the kidney or ureter, or acute hydronephrosis, pain on micturition is not infrequently caused by a pelvic abscess that lies close to the bladder or by an inflamed appendix that irritates the right ureter. Pain in the corresponding testicle may accompany renal colic, but testicular pain may occasionally occur with appendicitis (see Chapter 18) or ruptured abdominal aortic aneurysm. Pain accentuated by reclining and relieved by an upright posture is often of retroperitoneal origin, as in pancreatitis.

VOMITING

In acute abdominal lesions, apart from acute gastritis, vomiting is almost always due to one or both of these causes:

1. *Severe irritation of the nerves of the peritoneum or mesentery,* (e.g., consequent on the perforation of a gastric ulcer or of a gangrenous appendix, or torsion of an ovarian cyst pedicle).
2. *Obstruction of an involuntary muscular tube,* whether it be the bilary duct, the ureter, the uterine canal, the intestine, or the vermiform appendix.

1. Little imagination is needed to picture the intensity of stimulation of the nerve endings in the peritoneum by the acidic gastric juice flowing freely into the peritoneal cavity, nor is it surprising that a patient should vomit very soon after the onset of such irritation. But the rapid and copious pouring out of diluting fluid from the irritated peritoneal surface soon dilutes the acidic gastric juice and lessens the irritation, so that vomiting is seldom persistent after the perforation of a peptic ulcer.

In acute pancreatitis the celiac plexus is so intimately associated with the inflamed organ that the reflex stimulus is very great and vomiting is severe. It is frequently persistent, since for some time there is little mitigation of the severe stimulus apart from operation or gangrene of the organ.

Strangulation of a coil of intestine and torsion of the pedicle of an ovarian cyst are examples of sudden catastrophes in which sudden and severe stimulation of many sympathetic nerves causes vomiting to occur early and to be persistent.

2. Stretching of involuntary muscles causes pain, and if the stretching is extreme, vomiting occurs. Obstruction of any of the muscular tubes causes peristaltic contraction and consequent

stretching of the muscle wall, and vomiting is common. This is well seen in all the colics: renal, intestinal, and uterine. Behind the obstruction the tube becomes somewhat dilated, and as each peristaltic wave passes along, the tension and stretching of the muscular fibers are temporarily increased, so that the pain of colic comes usually in spasms. Vomiting usually occurs at the height of the pain. The term "biliary colic" is a misnomer, since colic designates a cramping or intermittency that is not characteristic of the pain of distention of the biliary tree. The steady nature of the pain of biliary distention is attributable to the lack of a strong muscular coat in the biliary tract.

In intestinal obstruction there is the additional factor that the contents of the intestine are mechanically prevented from progressing onward, and a reversed current may be set up, possibly as the result of antiperistalsis. In addition, absorption of salt and water is impaired when the intestine is distended and edematous.

3. It is well known that the stimulation of a particular nervous center in the medulla oblongata by certain drugs, such as apomorphine, may cause vomiting. It is possible that in intestinal obstruction or pancreatitis the vomiting may be partly due to the effect of absorbed toxins upon the medullary center. *The absence of vomiting is sufficiently common in a variety of abdominal catastrophes that the possibility of an acute surgical intra-abdominal problem should not be dismissed under these circumstances.*

The relationship of pain to vomiting. It is important always to inquire about the exact time of vomiting relative to the onset of pain. In sudden and severe stimulation of the peritoneum or mesentery vomiting is early, coming on soon after the pain. In actue obstruction of the ureter by a stone, or of the bile duct by a calculus, vomiting is early, sudden, and violent. In intestinal obstruction the length of the interval before vomiting ensues gives some indication as to how high in the gut the obstruction is situated. If the duodenum has become obstructed by a gallstone, vomiting comes on almost as soon as the pain, and is for a time, and until the stone passes on, frequent and violent. If the lower end of the ileum is occluded (apart from strangulation, which may lead to early vomiting), the vomiting may not occur for four or more hours, according to the acuteness of the stoppage. In large-bowel obstruction the vomiting is usually quite a late feature and sometimes may not occur at all, though nausea is often at a much earlier stage.

In appendicitis pain almost always precedes the vomiting, usually by three or four hours, sometimes by twelve or twenty-four hours or even longer. Rarely does the vomiting occur simultaneously with the onset of pain, and very rarely indeed does the pain start after the vomiting.

The frequency of the vomiting. This usually varies directly with the acuteness of the condition. Frequent vomiting *at the onset* of an attack of acute appendicitis usually means that the appendix is distended on the distal side of a stricture or concretion and signals the danger of possibly early perforation. The vomiting associated with acute pancreatitis is often frequent, occurring with the slightest provocation, but is scanty in amount.

There are, however, many serious abdominal conditions in which vomiting is either infrequent, slight, or absent. Internal hemorrhage from a ruptured ectopic gestation is often accompanied by little or no vomiting.

In obstruction of the large intestine vomiting is a late or infrequent symptom. This is very noticeable in cases of intussusception, where the absence of vomiting may obscure the acuteness and danger of the condition. When vomiting occurs in undoubted obstruction of the large intestine it is generally due to the incompetence of the ileocecal valve, allowing the back pressure to distend the lower ileum.

In obstruction high in the small intestine, vomiting is usually frequent and copious in quantity. Vomiting usually precedes pain in gastroenteritis.

The character of the vomit. This needs to be noted. In acute gastritis, which in severe cases may give rise to alarming abdominal symptoms, the vomit consists of the contents of the stomach possibly mixed with a little bile. In the colics the vomit is commonly bilious. In cases of acute torsion of a viscus, it is common for the patient to retch frequently but vomit very little. An infant with duodenal atresia will vomit bilious fluid, but the vomit in congenital hypertrophic pyloric stenosis will not contain any bile.

In intestinal obstruction the character of the vomited material varies. First the gastric contents, then bilious material, then greenish-yellow, yellow, and finally orange or brown feculent-smelling fluid is ejected. Feculent vomit is pathognomonic of obstruction of the *small intestine*—either mechanical or paralytic. Feculent vomit is rare in colonic obstruction.

Nausea and loss of appetite. These may be present in many patients who do not vomit. In the same person different grades of the same kind of stimulus may produce anorexia, nausea, or vomiting. Therefore it is always important to inquire about the presence or absence of all three of these symptoms. *Acute* loss of appetite is always significant. In a child especially, a sudden distaste for food accompanied by abdominal pain should always cause one to make a very careful examination for appendicitis. In questioning a child, it is well to find out the last meal that the child really enjoyed and the first for which there was no eagerness.

BOWELS

It is, of course, important to investigate the condition of the bowels, but it is unwise to attach too much importance to the simple fact of constipation unless accompanied by other symptoms. What is of greater importance is any significant departure from previously normal action of the bowel. In a person who is accustomed to a regular action of the bowels every day, the occurrence of constipation for several days may be of serious import, especially if accompanied by abdominal pain or flatulence.

The passage of several small, loose motions amounting almost to diarrhea is common at the onset of many cases of appendicitis in children. *When hypogastric pain and diarrhea are followed by hypogastric tenderness and constipation, suspect a pelvic abscess.* Tenesmus is sometimes a complaint in cases of pelvic abscess.

The presence of obvious blood or slime in the motions is to be questioned and looked for. The significance of blood and mucus in the rectum in the diagnosis of intussusception is well known.

MENSTRUATION

It is essential to inquire about the regularity of menstruation. It is not sufficient to ask whether the periods are regular; the exact date of the last period must be ascertained and any irregularity noted. Antedating or postdating of the normal period by a few days, or a longer or more profuse loss or the reverse, must be noted. Many patients with tubal gestation consider the bleeding that occurs at the time of threatening tubal abortion or rupture as the normal menstrual loss if the hemorrhage happens to coincide with the time of the usual monthly period. In such cases, however, it is nearly always possible by careful questioning to ascertain that the bleeding is in some way abnormal.

Pain accompanying the period of a woman who is not usually subject to dysmenorrhea should make one think of threatening early abortion, tubal gestation, or some condition unassociated with pregnancy. The timing of a ruptured follicular or corpus luteum cyst in relation to the menses is of obvious diagnostic importance.

History

It is well to inquire about any previous illness (e.g., previous operation, peritonitis, appendicitis, or pneumonia) that may possibly have a bearing on the present illness. One of the most important questions to ask the patient is: *Have you ever had a pain like this before?*

Since many if not most abdominal pains are loosely termed "indigestion," inquiry should be directed to the occurrence of any pain that has any relationship to the taking of food. Pain that comes on two or two and a half hours after taking food suggests duodenal ulcer. Constant epigastric pain made worse by the taking of food would make one suspicious of carcinoma or chronic ulcer of the stomach. Epigastric or right hypochondriac pain irregularly related to meals is in keeping with the presence of gallstones.

Previous attacks of jaundice, melena, hematemesis, and hematuria must briefly be inquired about, and it is well to find out if any great loss of weight has occurred.

Previous residence in tropical climes and any dysenteric history should be ascertained.

In a woman, previous confinements or pregnancies should be noted.

3. Method of diagnosis: the examination of the patient

The facial expression of the patient will on occasion furnish valuable evidence of the serious nature of the pain of which a complaint is made. When the nature of the pain is unclear because the patient finds it difficult to articulate its character, simple observation of recurrent discomfort on the face may tell the surgeon that the pain is intermittent and colicky. One should always take the time to watch the patient quietly for several minutes. I have often been told by the patient that the pain was steady, only to find this to be false by the simple expedient of careful inspection of the patient. The pale or livid face and sweating brow of a patient suffering from a perforated gastric ulcer, acute pancreatitis, or acute strangulation of gut is sufficiently distinctive, while the deathly pallor and gasping respiration of a woman with internal hemorrhage from rupture of a tubal gestation leave little doubt as to the diagnosis.

But appearances are often deceptive. Most surgeons of experience have seen patients suffering from perforation of a gastric ulcer whose complexion or facial expression gave no indication of the serious intra-abdominal condition from which they were suffering. In the majority of cases of early appendicitis, the facial appearance of the patient does not help at all. But in the late stages of all varieties of acute abdominal disease, the face tells the observer much that he ought to know but is sorry to learn. The dull gaze of the eyes and the ashen countenance of the patient suffering from severe toxemia, or the sunken cheeks and hollow-eyed aspect of the patient who has been vomiting repeatedly from internal obstruction or advanced peritonitis can always be recognized by the skilled clinician.

The back of the observer's hand placed on the patient's nose and cheek or on the extremities will often determine whether shock is present or collapse is impending, for a cool moist skin, due to failing capillary circulation, is usually indicative of one or the other. The presence of anemia, as shown by marked pallor of the cheeks, tongue, lips, and fingernails, should be noted even when the symptoms of sudden catastrophic hemorrhage are absent, for severe abdominal pain and some local tenderness may accompany the crises of the hemolytic anemias.

If the alae nasi are observed to be flaring, attention should always be particularly directed to the thorax, for a high temperature with moving alae nasi generally means pneumonia. It must be remembered, however, that any abdominal condition that in any way impedes the movement of the diaphragm *may* be accompanied by movement of the alae.

THE ATTITUDE IN BED

The attitude in bed is noteworthy. The restlessness of those suffering from severe colic contrasts with the immobility and dislike of movement evinced by the majority of those suffering from peritonitis. Ask a patient suffering from a perforated gastric ulcer to turn onto his side, and one sees at once the difficulty, circumspection, and pain with which the movement is accomplished. With extensive peritonitis the knees are frequently drawn up to relax the abdominal tension, while any inflammatory condition in contact with the psoas muscles causes a flexion of the corresponding thigh. Patients with pain of pancreatic or retroperitoneal origin very often prefer the sitting to the recumbent position.

In the early stages of many pathological states of the abdomen, however, no great help is obtainable from observing the attitude.

The pulse is too optimistic a friend to be relied upon for guidance, either in diagnosis or prognosis, in the early stages of acute abdominal disease. A normal pulse does not necessarily indicate a normal condition of the abdomen.

On the other hand, an increase in the frequency of the pulse is a constant accompaniment of the more advanced stages of peritonitis and hemorrhage, and after abdominal injury the careful observance of the pulse, hour by hour, is of great value in estimating the nature or seriousness of the intra-abdominal trouble.

The pulse of late peritonitis—small, hard, and so rapid as to be almost uncountable—is usually a terminal event, and is always of bad prognostic significance.

If acute abdominal disease is to be diagnosed in the early stages, it must first be recognized by the practitioner, who will see the patient at the beginning. Many, if not most, patients with serious acute lesions of the abdomen have a normal pulse during the early stages.

THE RESPIRATION RATE

This is chiefly of importance in differentiating between an abdominal and a thoracic condition. If the respiration rate is raised to double the normal *at the inception of the illness,* the causative lesion is probably thoracic in origin. In general peritonitis, however, or in cases of intestinal obstruction with great distention, or in severe intra-abdominal hemorrhage, the breathing may be much more hurried than usual, and with some persons nervousness produces shallow and more rapid respiration.

TEMPERATURE

One is more likely to obtain the correct temperature if the bulb of the thermometer is inserted into the mouth, under the tongue, or into the rectum. The oral temperature is likely to be spuriously low if the nasal passages are obstructed or if a nasogastric tube forces the patient to breathe through the mouth. If the validity of the oral temperature is in doubt, it should always be checked rectally.

A subnormal, normal, or raised temperature may accompany acute abdominal disease. All three may be recorded in the same patient at different times. In any severe abdominal shock or in severe toxemia, the thermometer frequently registers as low as 95°F or 96°F. This is the temperature, for example, often recorded at the *onset* of an attack caused by acute pancreatitis, acute intestinal strangulation, perforated gastric ulcer, or severe intraperitoneal hemorrhage.

At the onset of an attack of acute appendicitis the temperature is usually normal, but within a few hours it generally rises steadily to about 100°F or 101°F. When perforation occurs, the temperature usually goes a little higher.

A normal temperature is often seen in the early stages of a perforated gastric ulcer.

A ruptured ectopic gestation is usually accompanied by irregular, low-grade fever.

In intestinal obstruction the temperature is, as a rule, normal or subnormal. If any patient with abdominal pain is found to have a temperature of 104°F or 105°F at the onset of the illness, the thorax or the kidney is very likely the seat of the disease. High fever is quite unusual during the *early* stages of acute abdominal disease, except for severe pancreatitis.

ABDOMINAL EXAMINATION

Before examining the abdomen, it is well to learn from the patient the exact place where the pain started and if there has been any alteration in its location. The exact point of maximal pain should also be pointed out at the time of examination.

Inspection of the abdomen will reveal at a glance any abnormal local or general distention, and will in some cases determine the presence of a tumor or abdominal swelling. *All the hernial orifices must be inspected as a routine and special attention directed to the femoral canal, where, in an obese patient, a small hernia is easy to overlook.* Many lives are lost from failure to do this.

Movement on respiration. The respiratory movement of the abdominal wall is to be noted carefully, for any limitation indicates some rigidity of abdominal muscles. In the case of a perforated ulcer that has ruptured into the general peritoneal cavity, the abdominal wall hardly moves in any part on respiration, while in appendicitis with local peritonitis the hypogastric zone, especially the right iliac area, very frequently is immobile. In some cases of acute pancreatitis the epigastric zone may be motionless, and sometimes also the lower part of the abdominal wall. With biliary inflammation there is sometimes inhibition of the movement of the diaphragm, so that the respiratory movement of the epigastric area of the abdominal wall is diminished. In cases of pneumonia with pleuritic pain, one very salient finding is free movement of the abdominal wall on respiration, with restriction of the lower chest.

Palpation and percussion of the abdomen. In performing palpation of the abdomen, it is hardly necessary to remind the reader that the hands should be warm, that the examination should be commenced by the hand at that part of the abdomen farthest removed from the point of maximum tenderness, and that gentle pressure should be applied by the soft pulp of the fingers. *Gentleness is essential to success in palpation.* A rough, painful examination is not only difficult for the patient but may be misleading to the physician. Rough, painful palpation by the first examiner usually leads to fear on the part of the patient and possibly to misleading findings by subsequent examiners. Palpation determines the extent and intensity of the muscular defense or rigidity, locates any tender areas or hyperesthetic patches, and determines the presence of any swelling. It is well to have the patient's thighs flexed and supported in flexion while palpating the abdomen.

Percussion of the abdomen must always be carried out with the greatest gentleness. By its aid one can estimate the amount of distention of the gut, map out any dull area, estimate the extent of the stomach, determine abnormal liver dullness (see below), and sometimes exclude the presence of a distended urinary bladder. Its greatest value is to test for "rebound" pain (see below).

"Rebound" pain. If the fingers are pressed gently but deeply over an inflamed focus within the abdomen and then the pressure is suddenly released the patient may experience sudden, sometimes severe pain on the "rebound." *We do not recommend the performance of this test, for it elicits no more than can be ascertained by careful, gentle pressure and may cause unexpected and unnecessary pain.* Gentle percussion of the abdomen will provide more accurate information because it is in fact an easily regulated measure of rebound tenderness that can be more accurately localized.

Rigidity. Muscular rigidity is a relative term. The contraction of the muscles may be firm, continuous, and "like a board," as in many cases of general peritonitis due to perforated ulcer, or the muscular fibers may not contract to any detectable extent until the fingers are gently pressed on the abdominal wall, when the muscles spring to attention to defend the subjacent inflamed parts. In fact, a "boardlike" abdomen is less frequent than is generally believed, except in young individuals with a severe insult to the peritoneum. There is also an important mental factor in the production of abdominal wall rigidity. In some sensitive children, and in some very apprehensive adolescents and adults, the abdominal wall is held very rigid even though there is very slight intra-abdominal cause for it.

In pelvic inflammatory lesions and in intestinal obstruction, rigidity is often absent. *In rigidity due to thoracic disease, if the muscle contraction is overcome by continuous pressure the abdominal pain is not usually increased, but if there is a subjacent inflamed area in the abdomen the pain becomes greater as the hand presses in and overcomes the muscular resistance.* Some apprehensive patients have difficulty relaxing their abdominal muscles and prefer to make use of costal breathing even when there is no sufficient cause for abdominal rigidity. A simple way of encouraging them to use their abdominal muscles is as follows: A pillow is placed under the knees of the patient, and if need be, the chest may be supported by pillows. When about to palpate the abdomen with the right hand, the observer rests his left hand with a fair amount of pressure on the sternum of the patient, thereby inducing him to use the diaphragm and the abdominal muscles unless there is a serious lesion to prevent this.

It is exceedingly important to remember that muscular rigidity and resistance may be very slight even in the presence of serious peritonitis: (1) when the abdominal wall is very fat and flabby and the muscles are thin and weak; (2) when the patient is suffering from severe toxemia and the reflexes are dulled and diminished; (3) in elderly, feeble patients. *It is noteworthy that the recti are usually more definitely rigid than the lateral abdominal muscles. In elderly, feeble patients muscular rigidity is less noticeable.*

Hyperesthesia. Hyperesthesia may be tested by pinstroke, light pinch, or light stroking of the skin with a cotton applicator. It is wise to test the loin or lumbar region, as well as the thigh, for comparison with the findings on the abdominal wall. Although hyperesthesia along the distribution of the peripheral nerves whose terminals are irritated by the inflammatory process may be a helpful sign, especially in patients with acute appendicitis, its presence is so inconsistent as to be of relatively little diagnostic value in the vast majority of cases of *acute* abdominal pain. On the other hand, the presence of hyperesthesia may be a great help in patients with chronic or recurrent abdominal pain, by pointing strongly to nerve root irritation as the cause of the pain (e.g., root compression by protrusion of an intervertebral disc, spinal arthritis, or a tumor of the spinal cord).

A frequent corollary of the physical finding of hyperesthesia is the observation that the patient avoids wearing tight or constricting garments.

Unimanual and bimanual palpation of the loins. This is of help in detecting renal or other loin swellings.

The fingertips of one hand are pressed forward under the ribs of the corresponding side of the patient's body. Resistance and tenderness without swelling indicate rigidity and sensitiveness of the quadratus lumborum and adjacent muscles, due probably to a tender inflammatory focus nearby. A perinephric abscess, an inflamed kidney, or a retrocecal inflamed appendix may give this sign.

In bimanual palpation the observer brings the other hand to the front of the loin, and can thus feel between the two hands any loin swelling. A pyo- or hydronephrosis or a lumbar abscess can thus be detected. The patient should be asked to take a deep breath so that any movement of the swelling can be ascertained.

Determination of iliopsoas rigidity. It is well known that if there is an inflamed focus in relation to the psoas muscle, the corresponding thigh is often flexed by the patient to relieve the pain. A lesser degree of such contraction (and irritation) can be determined often by making the patient lie on the opposite side and extending the thigh on the affected side to the full extent. Pain will be caused by the maneuver if the psoas is rigid from either reflex or direct irritation (Fig. 5). The value of the test is diminished if the anterior abdominal wall is rigid. The psoas test is not so often positive when the inflammation becomes subacute.

Liver dullness. The estimation of liver dullness is occasionally of value. This is usually determined in the right vertical nipple line from the fifth rib to below the costal margin and from the seventh to the eleventh rib in the midaxillary line. If in a patient who has no signs or symptoms of an atrophic liver, and in whom there is no abdominal distention, a resonant note is obtained on percussing in

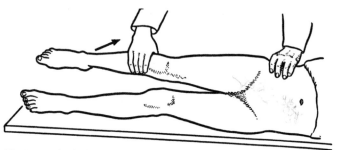

Fig. 5. Method of performing the iliopsoas test.

the normally dull liver area in the axillary line, then there is likely to be free gas in the peritoneal cavity due to the rupture of stomach or intestine.

The determination of liver dullness anteriorly is of no value in cases of great distention, for intestine may be pushed up in front of the liver and cause resonance that is higher than normal.

Free fluid. The determination of the existence of free fluid in the abdominal cavity by the proving of movable dullness is *not* of great importance in diagnosing acute abdominal disease. There are very few acute abdominal conditions in which there is not some free fluid, and in those cases where there is an easily estimable amount of fluid, there are usually other signs and symptoms that suffice. In the writer's experience it has seldom been necessary, and often inadvisable, to attempt to determine it. Free fluid in the abdomen may be serum, seropus, pus, blood, bile, or urine. If there is enough blood in the peritoneal cavity to cause movable dullness, hemorrhage should be evident from other symptoms. In peritonitis, with pouring out of a great deal of fluid, moving the patient causes pain, and the test is usually unnecessary.

In intestinal obstruction the determination of free fluid is of some value and causes the patient little inconvenience. One flank is percussed while the patient lies on the back and again after he has been turned over onto the opposite side. If the note changes from dull to resonant on changing position, free fluid is present. The test is sometimes vitiated by the accumulation of fluid within the coils of obstructed intestine. Such fluid is often great in amount and, owing to the atony and dilatation of the gut, moves easily from one part of the abdomen to another with a changed position of the patient.

EXAMINATION OF THE PELVIC CAVITY

As important as the examination of the main abdominal cavity is the investigation of the pelvis. The following methods should be employed in addition to the previous examination.

Suprapelvic palpation and percussion. This will have been done generally in the usual abdominal palpation, but it is well to pay special attention to the suprapelvic region. By pressing deeply behind one or the other Poupart's ligament, or behind the pubis, one may detect either deep tenderness, a mass, or muscular resistance indicative of deeper disease. A full bladder or an enlarged uterus, a high pelvic abscess, or an ovarian cyst may be thus discovered. If the bladder is full, further examination should be made after the bladder has been emptied.

Rectal examination. Rectal digital examination is extremely important and informative. The patient may lie on the side or the back. (In diffuse peritonitis or hemorrhage the patient may find it extremely uncomfortable to lie recumbent because of pain and dyspnea caused by diaphragmatic irritation.) A rubber glove or finger cot should be worn. The well-lubricated finger is gently introduced three or four inches up the rectal canal. By pressing forward, backward, upward, and laterally, the whole lower pelvis can be explored. Occasionally the finger cannot be passed up the canal on account of its being blocked by a large mass of inspissated feces; such a mass may cause chronic obstruction of the bowel and lead to more acute symptoms for which surgical assistance may be sought.

By pressing *forward* in the male, one can detect an enlarged prostate, a distended bladder, or diseased enlargement of the seminal vesicles. In the female one can palpate painful and painless swellings in Douglas' pouch, as well as enlargements and displacements of the uterus.

By passing the finger well up the canal, stricture of the rectum due to cancer or fibrosis, or ballooning of the canal below an obstruction can be ascertained. The apex of an intussusception may sometimes be felt.

It is important to test for tenderness on pressure against the pelvic peritoneum (see Fig. 19).

Bulging of a pelvic abscess against the anterior rectal wall can readily be detected. A Blumer's rectal shelf of carcinomatous implants in the cul de sac may be felt anteriorly and is sometimes acutely tender. It is often mistaken for a "high fecal impaction," a diagnosis that should rarely be made unless low impaction is also evident.

By pressing *laterally*, tenderness due to an inflamed and swollen appendix or a small abscess on the lateral wall of the pelvis can be elicited.

Posterior palpation will reveal any tumor or inflammatory mass on the pyriformis or in the hollow of the sacrum.

When the finger is withdrawn the presence on it of blood, slime, or pus should be noted.

In estimating the amount of pain caused by upward pressure on the pelvic peritoneum, one must not be misled by the patient's expression of general discomfort but must ascertain that the pain is due to pressure of the fingertip. Comparison of the discomfort produced by gentle pressure on the prostate or cervix versus that caused by pressure against the pelvic peritoneum may be extremely helpful in discerning the site of inflammation.

Bimanual rectoabdominal or vaginoabdominal examination will determine the presence and position of any pelvic tumor or swelling. It is especially important to note the size and position of the uterus. Any fullness in Douglas' pouch should be carefully palpated. A mass in the adnexa of the female may represent a tubal pregnancy, a tubo-ovarian abscess, or an ovarian tumor. A suitably situated appendiceal abscess may be indistinguishable from these conditions by physical examination, however.

In infants a bimanual rectoabdominal examination enables one to explore the lower part of the abdominal cavity thoroughly, and it may be possible to manipulate an intussusception between the fingers of the two hands. In patients with a colostomy or ileostomy, digital examination of the stoma should never be omitted because, as in infants, this procedure enables exploration of a surprisingly large part of the abdominal cavity.

Thigh-rotation test. When there is any inflamed mass adherent to the fascia over the obturator internus muscle, rotation of the flexed thigh, so as to put the irritated muscle through its full movements (especially internal rotation) will cause hypogastric pain. This sign should be tried especially when rectal examination is difficult or for any reason inadvisable (Fig. 6). The sign is positive when a perforated appendix, a local abscess, and occasionally a hematoma is in contact with the obturator internus or when there is an accumulation of inflammatory fluid in the pelvis.

EXAMINATION OF THE CHEST

The chest should be thoroughly examined by the usual methods of inspection, palpation, percussion, and especially auscultation. By this means diaphragmatic pleurisy, early pneumonia, and pleural effusion will be detected. The cardiac dullness must also be determined and the cardiac sounds auscultated.

SPINE

Any rigidity of or pain over the spinal column should be carefully observed, especially in children, in whom abdominal pain is frequently complained of during the course of spinal caries.

BLOOD PRESSURE

The estimation of the blood pressure is often of assistance in the diagnosis of acute abdominal crises. It is chiefly in cases of internal

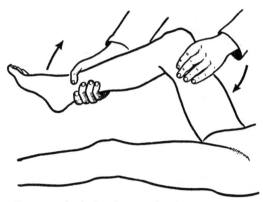

Fig. 6. Method of performing the obturator test.

hemorrhage, shock, and circulatory failure following intestinal obstruction that hypotension is present.

AUSCULTATION

It is generally held as a sacred aphorism that a quiet abdomen means peritonitis and that loud borborygmi indicate intestinal obstruction. While such is the case in classical examples, so many exceptions are found that undue weight must not be given to inconsistencies in auscultatory findings with the remainder of the clinical picture. *Rarely, if ever, should auscultatory findings themselves indicate or contraindicate an operation.*

Peristaltic sounds heard in the chest do *not* strongly suggest the diagnosis of a diaphragmatic hernia, and the presence or absence of sounds over an obvious bump in the groin cannot distinguish between an enlarged lymph node and an incarcerated femoral hernia. An abdominal bruit is common in the elderly and is diagnostic of neither aortic aneurysm nor vascular insufficiency to the intestine. *The absence of a bruit never excludes the presence of an aortic aneurysm.*

Of all the modalities of physical diagnosis of the abdomen, auscultation is one of the least valuable and most misleading. A friction rub is occasionally heard over an inflamed gallbladder and is of obvious import when noted over the lower thorax.

MEASUREMENT

Measurement of the abdomen has been used by some surgeons to give information as to the progress of a pathological lesion, particularly as to increase of distention. The tape must, of course, be placed around the abdomen at the same level each time for the comparison to be of value. Enormous distention may produce no more than a tiny change in the measurement, and the procedure seems of little value and is sometimes misleading. Simple digital estimation of the tension of the abdominal wall is more useful, especially if frequent comparisons are made by an interested and experienced observer.

4. Method of diagnosis: the grouping of symptoms and signs

In the two preceding chapters the methods of examination have been described. The next step is to piece the facts together in an approach to diagnosis. All abdominal crises give rise to one or more of five main symptoms or signs—*pain, shock, vomiting, muscular rigidity, or abdominal distention.* Most serious lesions present two or more of these main symptoms or signs, but sometimes pain may be the sole symptom. It may be of assistance, therefore, to consider the various symptom groups and the likely or possible causes of each group. The severity and the order of occurrence of the symptoms are important, and there are many other points (e.g., fever, diarrhea, constipation) which will be of help in differentiation.

What, then, are the main clinical pictures which the surgeon may encounter?

The first is:

ABDOMINAL PAIN BY ITSELF

A warning should be given here. Diagnosis would be so much simpler if pain and tenderness were always felt immediately over the site of disease, but such is not always the case. The initial pain of appendicitis is felt in the epigastric or umbilical region, while obstructive pain arising from the transverse colon is usually felt in the lower part of the abdomen (Fig. 7). The importance of the character of the pain, whether cramping, steady, sharp, burning, etc., cannot be overemphasized.

Abdominal pain is the only symptom in the earliest stage of quite a number of serious conditions. If the pain is felt only on one side of the abdomen, and if it radiates down to the corresponding groin or testicle, one must suspect renal colic and watch for blood in the urine. If the pain is felt all around the epigastric zone but seems greatest in the right hypochondrium and goes through to the subscapular region, one must suspect biliary colic.

CENTRAL ABDOMINAL PAIN

Suppose that the only symptom is severe central abdominal pain. What conditions must you bear in mind?

Severe epigastric or central abdominal pain may be due to simple intestinal colic, to the earliest stage of acute appendicitis, to the initial stage of obstruction of the small intestine, to acute pancreatitis, or even to the onset of mesenteric thrombosis. (Let it not be forgotten that the small-gut obstruction may arise in a small and almost invisible femoral hernia.) In addition to these intra-abdominal causes, the observer must remember that certain extra-abdominal conditions may cause deceptive pain within the abdomen; thus tabes dorsalis, herpes zoster, coronary thrombosis, and even acute glaucoma may temporarily mislead the practitioner (Fig. 8).

The golden rule in these cases is to examine the patient again within two or three hours. In every serious case there will be some other symptoms by that time, such as vomiting, fever, or local tenderness, which may point more definitely to the nature of the lesion.

In the case of appendicitis, after a few hours there will be nausea or vomiting, followed by right iliac tenderness and slight fever. *With a pelvic appendix, however, right iliac pain may be longer in appearing, but a rectal examination will reveal local tenderness.*

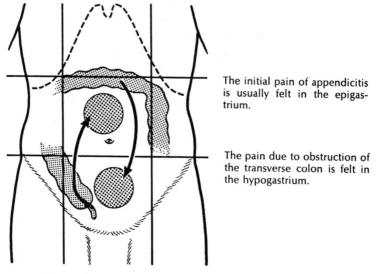

The initial pain of appendicitis is usually felt in the epigastrium.

The pain due to obstruction of the transverse colon is felt in the hypogastrium.

Fig. 7. The initial pain of a lesion is often felt far away from that lesion.

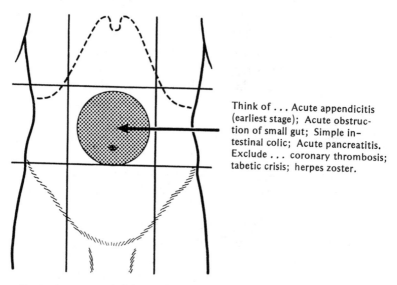

Think of ... Acute appendicitis (earliest stage); Acute obstruction of small gut; Simple intestinal colic; Acute pancreatitis. Exclude ... coronary thrombosis; tabetic crisis; herpes zoster.

Fig. 8. Acute central abdominal pain without any other symptom.

Simple intestinal colic, due to some indigestible article of diet or to incipient bacterial infection, is usually relieved by pressure, accompanied by audible gurgling. If the pain persists, something more serious should be suspected, for it is very common for patients with any kind of abdominal pain to attribute it to some recently ingested article of diet.

Tabes dorsalis, a rare condition, nowadays, may be excluded by testing the knee jerks and noting whether the pupils react normally to light and accommodation. With herpes zoster there will soon appear the characteristic rash, sometimes preceded by a zone of hyperalgesia, and with this condition there will be no nausea or vomiting. Pain may precede the development of the rash by several days, however.

The abdominal pain that sometimes accompanies acute glaucoma may distract both the patient's and the doctor's attention from the real source of the trouble. If this possibility is remembered, the mistake is soon remedied.

Another combination of symptoms is:

SEVERE CENTRAL PAIN WITH SEVERE SHOCK

When central pain is accompanied by collapse, pallor, subnormal temperature, and lowered blood pressure, yet without abdominal rigidity, the cause is more likely to be acute pancreatitis, mesenteric thrombosis, internal hemorrhage, ruptured aortic aneurysm, dissecting aneurysm, or even coronary thrombosis, and, in women, ectopic gestation.

Acute pancreatitis starts with severe epigastric pain, but pain is also felt deep in each loin and in the lower thoracic spine area. There may be some retching and, at a later stage, some epigastric rigidity. As the attack advances, irritative inflammatory fluid finds its way into the lower abdomen, particularly on the right side, so that the condition may be misdiagnosed as appendicitis. A history of attacks of biliary colic or alcoholism may put one on the right track, but one may be in doubt until a later stage when the serum amylase level may be found to be greatly increased or when Grey

Turner's sign is evident (discoloration in the loins). Shock may occur within an hour of onset but more commonly will be encountered six to eight or even twenty-four hours after pain begins.

Ruptured aortic aneurysm often begins with pain in the back, which frequently radiates into the groin, perineum, or scrotum. The pulses are not usually absent, and a pulsatile mass can usually be discerned.

Dissecting aneurysm starts with thoracic pain, which sometimes radiates down the left arm. Pain in the abdomen follows a little later and may lead to obliteration of one or both femoral pulses.

The early stage of *mesenteric thrombosis* may be very difficult to diagnose, but a previous history of intermittent claudication, abdominal angina, or other symptoms of serious arterial disease may give an indication of the true state of affairs.

Coronary thrombosis, which often causes severe epigastric pain, may lead to real difficulty in differentiation from an intra-abdominal lesion (see Chapter 24).

The pain and acute collapse from a severe hemorrhage into the peritoneal cavity (as may result from a ruptured aneurysm or a spontaneous rupture of the spleen in a man, or from the sudden rupture of an ectopic gestation in a woman) soon lead to an obvious anemia, which is accompanied by tumidity of the abdomen and sometimes by rapid, shallow respiration. These conditions are described elsewhere in this book.

PAIN WITH VOMITING AND INCREASING DISTENTION BUT NO RIGIDITY

This grouping usually means intestinal obstruction. A single initial vomit is common to nearly all urgent conditions within the abdomen, but repeated and persistent vomiting is indicative of obstruction of the small bowel or of spreading peritonitis; in the latter case, there is abdominal rigidity. With both conditions distention develops, but it comes on more quickly with peritonitis.

It is true that severe central pain and repeated vomiting without distention may be due to an acute gastritis, but before that diagnosis can be accepted, more serious lesions must be definitely excluded. Above all, one must exclude obstruction of the upper part of the small bowel. All the hernial orifices must be carefully examined, particularly the femoral rings. The abdominal wall must be examined for scars of former operations, and the vomit must be inspected to note any change of color from the bilious green to the dirty yellow that is the prelude to the stinking, feculent vomit of advanced obstruction of the small gut.

If the vomiting is not very frequent and distention of the ladder pattern gradually develops, one must expect the obstruction to be in the lower ileum. Visible peristalsis would, of course, aid the diagnosis. In any case of doubt, make another visit to the patient within an hour or two, and if possible have plain X-ray films of the abdomen in the supine and vertical positions made to see if there are any fluid levels in the small bowel or any significant distention of a coil of small intestine.

ABDOMINAL PAIN WITH CONSTIPATION, INCREASING
DISTENTION, AND PERHAPS VOMITING

Abdominal pain, accompanied by constipation and steadily increasing distention but little or no vomiting, is likely to be due to obstruction of the large bowel, particularly of the sigmoid colon. In a patient who has had abdominal pain, constipation, and increasing distention, if vomiting begins and becomes more frequent, then one should suspect obstruction of the large bowel with sudden failure of the ileocecal valve, with resulting back pressure into the lower ileum leading to symptoms of obstruction of the small bowel. A plain X-ray of the abdomen will give valuable information in these cases.

If abdominal distention of the large-bowel type develops rather quickly in an elderly patient, one must suspect a volvulus of the sigmoid colon. Sometimes a rectal examination may enable one to feel the thickened base of the twisted coil, and if an X-ray film can be taken, one will be able to see the distended coil quite clearly.

Before any operation for intestinal obstruction is undertaken, it should be the rule to examine the urine and to palpate the renal regions for any tumor. Uremia from any cause may lead to great abdominal distention and vomiting. Uremic distention may follow renal failure as a result of back pressure from an enlarged prostate, from bilateral polycystic disease of the kidneys, or from bilateral calculous pyonephrosis. On one occasion I have known the abdomen to be explored for supposed intestinal obstruction when the real cause of the distention and vomiting was renal failure due to lead poisoning.

Vomiting and distention in a newborn infant may be due to congenital obstruction of the small bowel. The diagnosis may be confirmed by an X-ray film, which will show fluid levels in the small intestine.

In children, one form of obstruction and one type of peritonitis (pneumococcal) need to be specially mentioned.

In infants of six to eighteen months, intussusception is fairly common (see Chapter 14). It occurs most often in healthy male infants. The symptoms commence suddenly: the infant screams, looks pale and ill, and draws up his legs. The pain subsides, the child seems better and may go to sleep, only to be awakened again by the recurrence of the pain, which causes another screaming attack. After several attacks of pain, examination may reveal a sausage-shaped swelling in the epigastrium or left side of the abdomen. Blood or mucus may be passed per anum or be seen on the finger following a rectal examination. A barium enema will clearly demonstrate an intussusception. (Pneumococcal peritonitis in children is discussed in Chapter 20.)

SEVERE ABDOMINAL PAIN WITH COLLAPSE AND GENERAL
RIGIDITY OF THE ABDOMINAL WALL

When a patient is found to be suffering from the above group of symptoms, the cause of the trouble is nearly always a visceral perforation, usually of the stomach or duodenum, consequent on the erosion of a peptic ulcer. Less common causes of general rigidity are perforation of the gallbladder or of a stercoral ulcer of the colon, and occasionally the rupture of a very distended appendix may quickly cause general rigidity and some collapse.

Of extra-abdominal conditions, bilateral pleuropneumonia may sometimes cause severe abdominal pain and rigidity; but if the rigidity can be overcome there will be no deep tenderness, and examination of the chest should reveal disease in that quarter.

With the gastric crisis of tabes, there is no rigidity of the abdominal wall unless there is associated peritonitis. It should not be forgotten that perforation of a gastric or duodenal ulcer *may* occur in a patient suffering from tabes dorsalis.

The symptoms of a perforated peptic ulcer are well known and characteristic. Severe general abdominal pain, often accompanied by initial collapse and always associated with a boardlike rigidity of the abdominal wall (Fig. 9), is followed in an hour or two by a reactionary stage in which the patient looks and feels better but the abdominal rigidity remains. This leads up to the final stage when peritonitis develops, the pulse rate increases, and vomiting and distention become prominent. There is often pain felt on top of the shoulder and the liver dullness may be diminished. If X-rays are available, air may be shown between the liver and the diaphragm (see Chapter 8 and Plate 2).

The group of symptoms that is perhaps most common and often the most difficult to diagnose is that in which pain, tenderness, and muscular rigidity are localized to one or another part of the abdomen. In most cases the diagnosis depends upon a careful discrimination among several possible causes. The various regions can only be considered briefly here.

Right hypochondriac pain and rigidity (Fig. 10). If the chest is clear, these symptoms are usually due either to an acute cholecystitis or to a leaking duodenal ulcer. (The differential diagnosis is more fully considered in Chapter 9.) It is sometimes difficult to differentiate between these two conditions before exploration. In tropical climes amebic hepatitis may cause similar symptoms, and in regions where hydatid disease of the liver is common, inflammation of such a diseased liver must be considered.

Left hypochondriac pain and rigidity. The left upper quadrant of the abdomen is least often the site of origin of local peritonitis, and

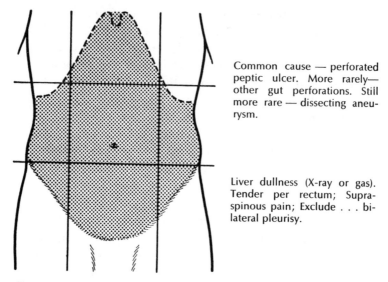

Common cause — perforated peptic ulcer. More rarely— other gut perforations. Still more rare — dissecting aneurysm.

Liver dullness (X-ray or gas). Tender per rectum; Supraspinous pain; Exclude . . . bilateral pleurisy.

Fig. 9. Severe abdominal pain with collapse and general rigidity of the abdominal wall.

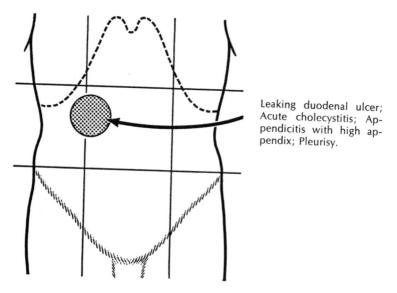

Leaking duodenal ulcer; Acute cholecystitis; Appendicitis with high appendix; Pleurisy.

Fig. 10. Tenderness and rigidity in the right hypochondrium.

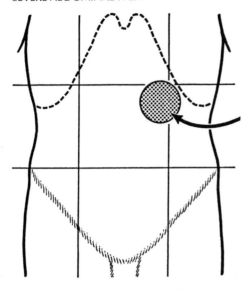

Pancreatitis; Perforated gastric ulcer (subphrenic abscess); Jejunal diverticulitis; Rupture of spleen (spontaneous); Leaking aneurysm of splenic artery; Acute perinephritis.

Fig. 11. Tenderness and rigidity in the left upper quadrant.

all the causes must be accounted rare. Although acute pancreatitis is undoubtedly the most common, one should consider the perforation of a gastric ulcer localized by adhesions, the rupture of an inflamed jejunal diverticulum, the leakage of an aneurysm of the splenic artery, and the spontaneous rupture of the spleen (Fig. 11).

In the last two conditions there will be evidence of hemorrhage; microscopic examination of a stained blood film might show evidence of leukemia. The history may be suggestive of gastric ulcer or of jejunal diverticulitis.

Right iliac pain, tenderness, and rigidity (Fig. 12). The most common cause of these symptoms is, of course, acute appendicitis, but there are many pitfalls, which are considered more fully in Chapter 6. Suffice to say here that a similar group of symptoms *may* result from disease of the pancreas, gallbladder, duodenum, right kidney, the ileocecal lymph nodes, a Meckel's diverticulum, the distal portion of the ileum (ileitis), a retained testis (in the male), and the right Fallopian tube and ovary (in the female). For differential diagnosis, see page 85.

Left iliac pain, tenderness, and sometimes rigidity. Diverticulitis of the left colon (Fig. 13). This combination of symptoms is most often due to an area of diverticulitis in the iliac or sigmoid colon, but other causes must be considered.

The descending, iliac, and sigmoid colons contain solid or semi-solid material, and after middle age there may develop small recesses or pouches, often surrounded by fat, in which particles of feces may collect and stagnate. These diverticula may be very numerous, often get inflamed and cause thickening of the wall of the colon, which sometimes forms a palpable mass in the left iliac fossa or the pelvis.

Acute or subacute inflammation of a segment of the left colon affected by diverticulosis may spread to the contiguous peritoneum

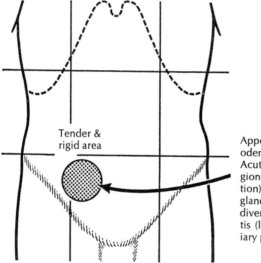

Tender & rigid area

Appendicitis; Leaking duodenal ulcer; Pyelitis; Acute pancreatitis; Regional ileitis (exacerbation); Inflamed ileocecal glands; Inflamed Meckel's diverticulum; Cholecystitis (low gallbladder); Biliary peritonitis.

Fig. 12. Tenderness and rigidity in the right iliac region.

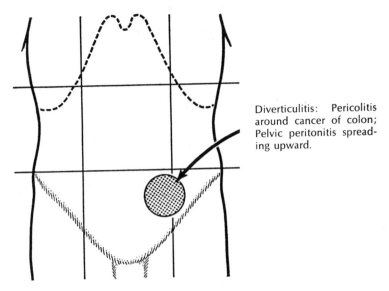

Diverticulitis: Pericolitis around cancer of colon; Pelvic peritonitis spreading upward.

Fig. 13. Tenderness and rigidity in the left iliac region.

and retroperitoneal space, and occasionally an abscess forms and may burst into the bladder or make its way to the skin surface.

The symptoms of such local inflammation are local pain in the left iliac region, tenderness on pressure, and sometimes rigidity of the overlying muscular abdominal wall. When the condition subsides, a lump may be felt in the iliac fossa. During the attack there may be slight fever, as well as initial epigastric pain, but vomiting is not common. The symptoms resemble closely those that would be manifested by a retrocecal, left-sided, acutely inflamed vermiform appendix, but the appendix is very rarely found on the left side.

Pain and tenderness in the left iliac region may also be due to inflammation around a cancer of the iliac colon. The same symptoms may arise from Crohn's colitis, which can be indistinguishable from diverticulitis. It must also be remembered that the inflammatory symptoms due to a perforated pelvic appendix may ascend from the pelvis to the left iliac fossa. In addition, the observer must be on the lookout for an acute pyelitis with a low-lying left kidney, and must remember that a left-sided pleurisy may cause pain and possibly rigidity in the left iliac region.

Hypogastric pain and rigidity. Hypogastric pain and rigidity in a young or middle-aged man is usually due to a *perforated, inflamed appendix;* in an older man, an alternative diagnosis would be a *perforated sigmoid diverticulum.* The same symptoms in a young woman might be due either to a diseased appendix or to a pathological condition of the uterus, fallopian tubes, or ovaries. The differential diagnosis is considered more fully in Chapter 7.

5. Laboratory and radiological tests

In its many editions this textbook has always emphasized the *clinical* evaluation of the patient with abdominal pain, and it continues to do so. This chapter has been written mainly to offer a perspective on the judicious use of the plethora of tests currently available to the clinician. *Overreliance on laboratory tests and radiological evaluations will very often mislead the clinician, especially if the history and physical examination are less than diligent and complete.*

Nevertheless, these tests can clearly be useful if the physician or surgeon selects them carefully and understands their pitfalls and limitations. This is especially true if the physician caring for the patient is directly involved in the evaluation and interpretation of the test results. No specialist is in a better position to interpret these tests than the doctor who is directly responsible for the care of the patient.

In many medical centers, there are standard algorithms for laboratory and radiological investigations. As a result, not only do patients with abdominal pain receive many unnecessary, expensive, and sometimes misleading laboratory tests, but the more important history and physical examination are neglected. It is imperative that the clinician understand that *laboratory tests will never substitute for clinical evaluation of the patient.* It would require far too many pages to recount all the ways that mistakes have been made through rigid adherence to the outcome of laboratory tests. Unfortunately, there has grown up a mistaken belief that a test result which arrives in the form of numbers or a report is somehow more reliable than a clinical evaluation—nothing which could be further from the truth! *In many instances the decision to operate must be made without laboratory confirmation of a diagnosis.*

STANDARD X-RAYS OF THE ABDOMEN (PLAIN AND UPRIGHT)

X-ray studies are *not* necessary or desirable in all patients with abdominal pain. In patients with a clear diagnosis of strangulated inguinal or femoral hernia, there is no justification for films. The same can be said for cases of easily discernible acute appendicitis, cholecystitis, perforated ulcer, or involvement of the female genital organs.

When the cause of right-sided abdominal pain is obscure, however, the radiological finding of a calcified fecolith in the right lower quadrant sometimes may help establish the diagnosis of acute appendicitis so that the appropriate incision may be made (Plate 3). Radiopaque gallstones (15 percent of cases) or gas bubbles in the wall or lumen of the gallbladder in a patient with symptoms suggestive of acute cholecystitis will almost certainly corroborate that diagnosis, although a *palpable*, tender gallbladder by physical examination in such instances is equally reliable. Free air in the peritoneal cavity will help the surgeon diagnose a perforated ulcer in doubtful cases of right upper quadrant pain. Remember, however, that free air cannot be detected in as many as 20 percent of cases of perforated ulcer. Free air can easily be missed on the film when it is visible only between loops of intestine. For the best opportunity to observe free air under the diaphragm or the abdominal wall, where it is most readily detected, the patient should be kept in the semiupright or lateral decubitus position for at least five to ten minutes before the upright or decubitus film is taken, a fact often ignored by technologists in radiology departments.

In patients with simple, nonstrangulating obstruction of the small intestine and in some with colonic obstruction, the plain film may be diagnostic. But it must be remembered that in at least 50 percent of patients with strangulation obstruction of the small intestine, plain films will be *normal* or nondiagnostic.

CONTRAST X-RAYS

If contrast studies are required, water-soluble contrast media are generally used when perforation of the gastrointestinal tract is strongly suspected because of the adverse effects of barium in the peritoneal cavity. These water-soluble agents do not provide high-quality results of the type that may sometimes be required to delineate obstruction of the small intestine or colon, but they are usually adequate to reveal perforation. The widespread use of water-soluble media to study *all* patients with acute abdominal pain who require contrast studies, a policy practiced in many institutions, is to be condemned.

In certain instances of suspected perforation of a gastric or duodenal ulcer, a contrast study with water-soluble agents may be useful. For example, the demonstration of leakage is diagnostic in doubtful cases; and in late cases suggestive of continued leakage, the failure to demonstrate extravasation of dye together with the absence of *generalized* peritonitis by physical examination usually calls for nonoperative treatment.

In some patients with partial obstruction of the small intestine, the plain X-ray film is equivocal, especially if it is not taken during the height of an attack. Thus, the diagnosis of stenosis or malignancy of the small intestine can easily be missed unless a contrast meal with *barium* is given during an acute episode. There need be no fear of the use of barium in these cases because the content of the small intestine is completely liquid, and impaction of barium in the small intestine will not occur. Contrarily, barium should never be given *by mouth* when obstruction of the *colon* is suspected because an incomplete obstruction can be converted into a complete one by the barium.

If the history, examination, and plain X-ray indicate that colonic obstruction is present, a contrast enema should be done, usually with barium. This study may indicate the exact point and causation of the obstruction and also may serve to reduce an intussusception or untwist a volvulus. It is a matter of judgment in the individual clinical situation whether contrast enema or endoscopy is used first, since the two procedures are usually required and are not mutually exclusive. Should the question arise of perforated diverticulitis, free perforation of the colon, or colonic anastomosis into the peritoneal cavity, it is better to demonstrate such a perforation with water-soluble contrast than to procrastinate for fear of harm from the study.

Intravenous pyelograms are needed in cases of suspected ureteral colic when a plain X-ray film fails to show the calculus (15–20 percent of the cases). In addition, they may be of particular value when acute pyonephrosis due to ureteral obstruction is suspected. If a seriously injured patient is thought to have a renal injury, a rapid intravenous pyelogram should be carried out (on the operating table, if necessary) to demonstrate whether a normal, uninjured contralateral kidney is present, in case the need arises to remove an irrevocably damaged kidney. Angiograms of the splanchnic vasculature should be carried out in patients with suspected occlusive or nonocclusive ischemia of the intestine (see Chapter 16).

NUCLEAR SCANS

Intravenous cholangiography has largely been replaced in recent years by radioisotopic scans that employ material (radioactive technetium) that is readily excreted by the normal liver and that can easily visualize the normal gallbladder and extrahepatic bile ducts. If the gallbladder is visualized, the diagnosis of acute cholecystitis can be excluded from consideration. However, the converse is far from true. Many seriously ill patients, especially in circumstances where intrahepatic cholestasis is common, may show no demonstrable radioisotope in the gallbladder; hence the study can be misleading. The presence of a palpable, tender, distended gallbladder is generally of greater diagnostic value than these tests, and it should indicate that these studies are *not* necessary.

Such scans can be of value in differentiating acute pancreatitis from acute cholecystitis since the isotope should be visualized in the gallbladder in most cases of acute pancreatitis. But in extremely ill patients with severe pancreatitis, the gallbladder cannot always be visualized.

CHEST X-RAYS

In some hospitals, especially in the United States, chest X-rays are routine for all patients with abdominal pain, a practice whose costs far outweigh its benefits. However, if a question of pneumonia or pulmonary infarction has arisen, these studies are mandatory. Occasionally, free air beneath the diaphragm is detected by chest X-ray when it is not seen in abdominal upright or decubitus films. This usually means that the patient was not kept in an upright position for a sufficient length of time before the film of the abdomen was made.

Pleural effusions on either side are very common in patients with subdiaphragmatic inflammatory processes, including acute diffuse peritonitis. A left-sided pleural effusion in a patient with central epigastric and left upper quadrant pain usually means pancreatitis, whereas a right-sided effusion is more likely with perforated ulcer.

It has become a common practice to tap such effusions, especially in nonsurgical departments. I have seen one patient who developed a rapidly progressive, fatal empyema and another who almost died because the needle seeking a small pleural effusion traversed either the infected peritoneal fluid or the splenic flexure below. These pleural effusions are almost invariably sterile and small in amount. Unless one is seeking an elevated amylase level in a patient with suspected pancreatitis, there is little to be learned from thoracentesis. The dictum that virtually all pleural effusions require tapping for diagnostic purposes can be extremely dangerous.

An elevated, immobile diaphragm may be a helpful finding since it usually means that there is a subdiaphragmatic inflammatory process of some kind.

ULTRASOUND AND CT

Many physicians now appear to believe that unless a clinically palpable mass can be confirmed by ultrasonography or by CT scan, it simply does not exist. Yet as we evaluate all laboratory tests, it becomes increasingly evident that clinical findings are often more valid than test results if clinical skills are fully matured.

Ultrasound can be of value in *selected* patients with *acute* abdominal pain. With suspected acute cholecystitis, it is not needed if the gallbladder is clinically palpable. When the gallbladder is not felt, the ultrasonic demonstration of gallstones may be useful. Remember, however, that the difference in size between a normal gallbladder in the fasting state and a tensely distended, acutely inflamed gallbladder of acute cholecystitis is small and may be indiscernible. Occasionally, an enlarged pancreas consistent with pancreatitis can be detected by ultrasound, but often the gas in overlying distended intestine interferes with this study.

Recently, some have suggested that visualization of a thick-walled appendix by ultrasound is diagnostic of appendicitis because the normal appendix usually cannot be discerned. This reasoning is based on flawed studies. At present, *the best use of ultrasound in a confusing case of possible appendicitis is to exclude other lesions,* especially in the female (e.g., tubal pregnancy, pyosalpinx).

Cystic collections almost anywhere in the abdomen can be outlined ultrasonographically, but deciding the nature of the collection is in the hands of the clinician. Ultrasound can be especially helpful in patients with subdiaphragmatic abscess or other postoperative collections, or in obese patients where a cystic lesion in the pelvis may be difficult to define on bimanual examination. In young girls with a dermoid of the ovary, ultrasonography can avert the need to use diagnostic techniques that require radiation. Aneurysms of the aorta also lend themselves to detection by this method.

I have encountered several situations in which large, ultrasonically demonstrated pelvic fluid collections were mistaken for the urinary bladder, leading to unnecessary cystoscopies and intravenous pyelograms; conversely, on several occasions the urinary bladder has been mistaken for a mass. CT and magnetic resonance imaging (MRI) are expensive and have a limited place in the study of patients with *acute* abdominal pain. Although some authors have recently suggested that patients with blunt trauma should be subjected to CT scans because ruptures of the liver, spleen, and pancreas are readily demonstrated (and indeed they are), the time needed to obtain such examinations in seriously injured and shocked patients is often excessive in most institutions. However, because of differences in the degree of embryological rotation of the cecum, the appendix may lie in any position, even in the left upper quadrant! Ruptures of the spleen in hemodynamically stable patients can also be nicely shown by radioisotopic scans of that organ. CT scans are of greatest value in detecting hepatic abscesses or tumors and in confirming the ultrasonically detected fluid collections or aneurysms mentioned above. Their main use is in patients with *chronic* rather than acute abdominal problems.

ENDOSCOPY

Upper gastrointestinal endoscopy has virtually no place in the diagnosis of acute abdominal pain. On the other hand, flexible sigmoidoscopy and colonoscopy are extremely useful in patients with colonic obstruction, as they often reveal the cause of the obstruction, whether carcinoma or diverticulitis. Polyps that cause intussusception can be removed and volvulus of the sigmoid colon can be untwisted. In patients with unrelenting, adynamic ileus of the colon, decompression of the large intestine can sometimes be achieved with the colonoscope, thus averting the need for operative decompression.

ABDOMINAL PARACENTESIS

Abdominal paracentesis with a No. 20 or No. 21 needle can be very useful in some cases of peritonitis of obscure origin. For example, analysis of the fluid obtained may help establish the diagnosis of primary pneumococcal peritonitis, or the *gross* characteristics of the fluid may distinguish between a perforated ulcer and acute pancreatitis. It is to be cautioned, however, that the amylase level of the fluid may be elevated in perforated ulcer and strangulation obstruction as well as in acute pancreatitis. I prefer a relatively short needle inserted into the peritoneal cavity after infiltration of the abdominal wall with a local anesthetic. The procedure should not be done in patients with a tensely distended abdomen, and *the region of the gallbladder* should be avoided (Fig. 14). The gallbladder does not vigorously contract around a puncture site, as does the intestine, so that bile peritonitis may result if the needle traverses the wall of the gallbladder. If the needle enters the intestine when the abdomen is grossly distended, a remarkably common occurrence under such circumstances, obtaining feculent material may lead to the false diagnosis of fecal peritonitis.

Puncture of the peritoneal cavity is of greatest value in patients who suffer blunt abdominal injury, especially if they have also sustained an intracranial injury that renders them unconscious. Such procedures are most safely accomplished by inserting a long plastic intravenous cannula or peritoneal dialysis catheter after a small incision through the linea alba down to the peritoneum is made midway between the pubis and the umbilicus. The catheter is inserted through the peritoneum under direct vision with suitable local anesthesia, but only after the operator has made certain that the bladder has been emptied by means of a catheter. One liter of isotonic saline can be infused into the peritoneal cavity and then allowed to return to a collection bottle. Fluid sufficiently opaque to preclude reading newspaper print through it is a signal for exploration. Somewhere between 50,000 and 100,000 cells/mm^3 is the level at which intervention is required. When a patient with blunt

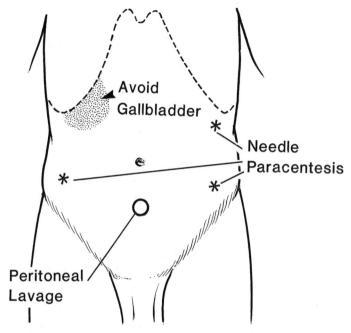

Fig. 14. Proper sites for needle paracentesis or peritoneal lavage.

trauma is sufficiently conscious to show by physical examination
that peritonitis is present, this procedure is unnecessary.

LAPAROSCOPY

This procedure has been strongly advocated by gynecologists to aid
in the diagnosis of ectopic pregnancy and ovarian cysts and to
distinguish pelvic inflammatory disease from acute appendicitis. It
requires a general anesthetic and may be very much restricted by
the presence of adhesions from previous operations. Unless the
entire appendix can be visualized, a pelvic purulent exudate may
be erroneously attributed to salpingitis when in fact a ruptured
appendix is present. If there is any question about the pathological
process, a full open laparotomy should be done. Laparoscopy, in

my opinion, is an extremely limited and difficult way to explore the abdominal cavity. There are no well-controlled data at present to indicate that laparoscopy has great advantages over laparotomy when a knowledgeable clinician deems the latter to be required.

STANDARD BLOOD AND URINE TESTS

Excessive reliance upon blood tests often leads physicians astray. For example, many patients with acute appendicitis have a normal white blood count, even after perforation. To expect a patient with acute intra-abdominal hemorrhage to show anemia at an early stage is to delay unnecessarily and thus endanger the patient's life. But hemoconcentration and abnormal serum electrolytes and blood urea nitrogen in a patient with a perforated ulcer or intestinal obstruction may provide a useful quantitative measure of the fluid therapy that will be necessary for resuscitation before operation. In cases of suspected diabetes, the level of blood glucose will, of course, be invaluable.

It cannot be overemphasized that the responsible physician on occasion must personally examine the blood smear should the question of blood dyscrasia arise and must personally study the sediment of the urine when ureteral colic or a urinary infection is suspected. Even a small number of red blood cells in the urinary sediment may help establish the diagnosis of ureteral stone in a patient with a compatible history and findings.

In many emergency units, I have encountered the pernicious practice of ordering a large battery of laboratory tests and X-rays, often by a triage nurse, before the patient is examined by a physician. Not only have seriously ill patients thus sustained considerable delay and suddenly gone into shock in a radiological unit, but the return of "normal" reports has often lulled emergency room personnel into believing that all is well and that no serious intra-abdominal condition exists. *Physicians should always take a thorough history and perform a physical examination before any tests are done.* Only if this rule is followed will the proper tests be ordered.

6. Appendicitis

It is essential that the earliest signs and symptoms of appendicitis should be appreciated clearly; the view is accepted by most surgeons of experience that every patient with acute appendicitis should be operated on within the first twenty-four hours from the onset or as soon thereafter as possible. There are two reasons why cases come to attention later than this: either the patient may think that the symptoms are not serious enough to need medical advice, or the medical adviser may think the symptoms not typical of appendicitis or not serious enough to demand operation. The remedy for the first is the education of the public, and for the second to remind the reader that the so-called typical symptoms of appendicitis as given in the textbooks sometimes indicate a somewhat advanced stage of the condition, and that it is impossible to say at the beginning of an attack whether it is likely to be mild or severe in type.

It is desirable, and in many cases possible, to diagnose appendicitis before peritonitis has set in, or at least before there is any more than that slight amount of irritation of the peritoneum that is commonly associated with any inflammatory process within the gut.

Pathological condition in relation to symptoms. The different grades of inflammation of the vermiform appendix have for many years been well described and understood, though there is still a lack of appreciation of the advanced pathological condition often coexistent with the initial symptoms or, at any rate, with the initial complaint. Mild inflammation of the mucous membrane, parenchymatous inflammation of the whole wall, gangrene of the interior lining or of all but the peritoneal coat—any of these may coexist with symptoms so slight (but *not* indefinite) that they may be overlooked by the patient and thought to be of slight significance by the unobservant onlooker. Even rupture of the appendix due to local gangrene may not cause the patient to become bedridden or seek medical advice, so long as local adhesions prevent the extension of the mischief. When the appendix ruptures into the general abdominal cavity in the absence of any protective adhesion, or when, after being localized, the inflammatory process extends, not even the most stoical or insensitive patient can refrain from seeking advice and taking to bed.

Obstruction of the lumen of the appendix, either by a concretion, stricture, kink, or adhesion, is usually accompanied by more acute and severe symptoms. So definite is the difference that some describe two forms of the disease—acute appendicitis and acute appendical obstruction. So far as this differentiation tends to emphasize the usually greater urgency of symptoms with obstruction of the lumen of the appendix it serves a useful purpose, but to treat the obstructive form as a different disease is unnecessary.

When the appendix has ruptured, the pathological condition is not merely appendicitis but peritonitis—local, diffuse, or general, as the case may be—and diagnosis is to that extent more complicated. It is frequently only by careful attention to the history that one can make certain as to the true cause of such peritonitis.

Anatomical position of the appendix in relation to symptoms. The vermiform appendix, though usually described as being situated behind the ileocecal junction, with the tip directed toward the spleen, is not always found in that situation when it is diseased and sought by the surgeon. To know the common positions is of great importance in diagnosis, for the signs and symptoms may thereby vary considerably. The accompanying diagram shows the more common positions (Fig. 15). For descriptive purposes it is well to recognize the ascending appendix, the iliac appendix, and the pelvic appendix. However, because of differences in degree of embryological rotation of the cecum, the appendix might lie in any position, even in the left upper quadrant! When the appendix lies by the side of the ascending colon, or in the iliac fossa, there will be the most definite local signs, while if it is situated behind the cecum, or behind the distal portion of the ileum and the common mesentery, the inflammatory process will be somewhat masked by the gut lying in front. If the appendix hangs over the right brim of the true pelvis, the disease may give rise to few signs in the suprapubic region of the abdomen, and a dangerous condition results to which we shall call attention below.

Many of the mistakes made in the diagnosis of appendicitis are due to a failure to realize the very great difference in signs and symptoms that follow from the varying positions and relations of the appendix.

Diagnosis of appendicitis before perforation has taken place

In any case of suspected appendicitis one must consider carefully:

1. The history immediately prior to the onset of pain.
2. The symptoms of the attack and the local signs.
3. The *order of occurrence* of the symptoms.

The local conditions are more variable and notable after perforation has taken place, but there are definite indications even before perforation.

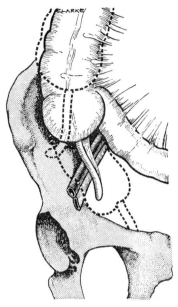

Fig. 15. The various positions the cecum may occupy.

THE HISTORY

There is frequently a *history* of indigestion, "gastritis," or flat-ulence for a few hours and sometimes a day prior to the onset of the attack. In a patient who has never, or seldom, been subject to pain after taking food, this history should be sufficient to put one on guard. It may be elicited that frequent slight attacks of pain have been experienced in the appendicular region.

A history of unusual irregularity of the bowels is often obtained. Sometimes there is constipation, at other times diarrhea, especially in children. The occurrence of diarrhea is likely to mislead. In some cases an inflamed pelvic appendix may irritate the rectum by contiguity. This early diarrhea has to be distinguished from the late variety due to irritation of the rectum by pelvic peritonitis or a pelvic abscess. The latter is manifested by frequent small movements, often consisting mainly of mucus and without truly voluminous diarrhea.

THE SYMPTOMS AND LOCAL SIGNS OF THE ATTACK

The signs and symptoms (Fig. 16) are:

1. Pain (epigastric, then right iliac).
2. Vomiting, nausea, acute loss of appetite.
3. Local tenderness (per abdomen or per rectum).
4. Local rigidity of muscles (inconstant).
5. Local distention (inconstant).
6. Superficial hyperesthesia (inconstant).
7. Fever (inconstant).
8. Constipation (inconstant).
9. Testicular symptoms (uncommon).

Pain. The pain in the majority of cases is first referred to the epigastric or umbilical region and only later is localized in the right iliac fossa. Occasionally the initial pain is felt "all over the abdomen," though that is more usual in cases with perforation. Sometimes the pain is from the first hypogastric. The initial pain is usually vague, resembling excessive flatulence or indigestion and very frequently giving the sensation that passage of stool or flatus might relieve the discomfort—the so-called downward urge. The downward urge is a very common phenomenon, but even if stool or gas is expelled, the discomfort remains unrelieved. When the appendix is retrocecal in position the initial pain may be felt in the right iliac region. Therefore, though the pain commonly starts in the upper or central part of the abdomen, this is not invariable.

The early pain of appendicitis is visceral rather than somatic in origin and is thus a referred type of poorly localized discomfort. First and most important is the exaggerated peristalsis of the appendix, excited by the relative or absolute obstruction to its

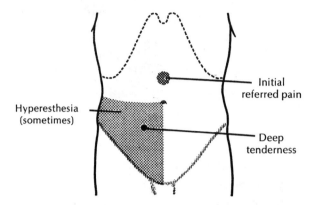

Initial
referred pain

Hyperesthesia
(sometimes)

Deep
tenderness

Fig. 16. The common position of initial referred pain, the position of deep tenderness (nearly always to be elicited when the abdominal wall is not rigid), and the iliac triangle of hyperesthesia found in some cases of appendicitis.

lumen by a concretion, kink, or swollen mucous membrane; bacterial infection causes the accumulation of irritating products, which leads to a distention of the appendical lumen.

The epigastric pain is most acute and distinct in those cases where there is considerable obstruction to the appendical lumen. *The localization of the pain to the right iliac region usually takes place some hours after the onset of the diffuse pain in the epigastric or umbilical region.*

Vomiting, nausea, and anorexia. Vomiting generally occurs in the early stages of the attack, but usually a few hours after the initial pain. Many patients do not vomit, but instead have a sensation of nausea. Loss of appetite or repulsion for food may be regarded as a lesser degree of the same sensation and often of equal value in diagnosis. Anyone in previously good health who suddenly loses appetite and complains of abdominal pain should be carefully watched for appendicitis. The degree of nausea and the frequency of vomiting in the early stages appear to depend on the amount of distention of the inflamed appendix. Vomiting is more prone to occur in children or in patients whose digestive tract is easily deranged. So frequent is anorexia or nausea, at least to some degree, that the presence of hunger should raise a serious question of the diagnosis of acute appendicitis.

It may be taken as a general rule that the severity and frequency of the vomiting at the onset of an attack of appendicitis indicate the degree of distention of the appendix and consequently the immediate risk that perforation may occur. Vomiting before the onset of pain is extremely rare in acute appendicitis and should lead to suspicion of a different diagnosis.

Local tenderness. Local tenderness over the site of the appendix is frequently absent at the onset of the attack, and for some time the local symptoms are masked by the more general abdominal pains. When the latter have subsided, the local tenderness is usually easily elicited. Occasionally, even careful palpation cannot detect any spot of local tenderness in the iliac fossa during the *initial* stage of appendicitis. Early deep tenderness can most often be detected just below the middle of a line joining the anterior superior iliac spine and the umbilicus, corresponding roughly to the position of the base of the appendix. Tenderness over McBurney's spot is not so constant. This tenderness appears to be located in the appendix itself, for the site of the pain on pressure varies somewhat according to the position of the appendix and is obtainable when that viscus is not adherent to any surrounding part. Sometimes the tenderness may be due to adjacent peritoneal irritation. The spot of maximum tenderness may sometimes be accurately located by gentle percussion over the affected region. In the case of an appendix situated in the pelvis, a rectal examination will frequently elicit pain when the finger presses on the inflamed organ, while at the same time abdominal tenderness may be absent. Tenderness of the abdominal wall is diminished in cases in which the inflamed appendix lies behind the cecum or behind many coils of small intestine.

Whatever the constellation of signs and symptoms, the clinical diagnosis of acute appendicitis cannot be made unless tenderness (no matter how slight) can be demonstrated *in some location*.

Hyperesthesia. Local hyperesthesia of the skin of the abdominal wall is an occasional but inconstant accompaniment of an inflamed, unperforated appendix. It is usually confined to the right side. The areas affected nearly always lie in the area of distribution of the nerves from the tenth, eleventh, and twelfth dorsal and first lumbar spinal segments.

Rigidity. Local muscular rigidity over the inflamed area is frequently present but is by no means a constant finding in the initial stages. There are several grades of muscular rigidity. The extreme degree is that in which the particular section of the abdominal wall is persistently stiff and will not move on respiration; in a lesser degree the muscle stiffens almost as soon as the hand touches the skin, and in the least degree the rigidity occurs only (and that to a slighter degree) when the fingers are pressed more deeply into the iliac fossa or toward the appendix. In most cases extreme muscular rigidity coincides with commencing peritonitis, and even slight degrees, when persisting, are due to irritation of the parietal peritoneum. It is a common experience to find no local muscular rigidity in a case of appendicitis without any peritonitis. In making this statement, it must be understood that great care should be taken to exclude the rigidity that many patients develop as a result of nervousness and apprehension, or that may be induced by a rough or cold examining hand. Certainly, with an unperforated appendix situated in the pelvis, rigidity of the abdominal wall is nearly always absent. Failure to realize this important fact is responsible for many delayed operations and lost lives. *An appendix may be on the point of bursting into the general peritoneal cavity without a single adhesion to limit infection, though at the same time the abdominal wall may be flaccid and allow free manipulation without any rigidity.* This fact must be known to every surgeon of experience, but in general it is certainly not fully appreciated. Rigidity is taught to be one of the earliest signs of acute inflammation of the appendix, whereas in quite a large proportion of cases it is almost completely absent in the earliest stage, and in some cases it is absent even though pelvic peritonitis exists. When the inflammation of the appendix has caused edema of the contiguous portions of the abdominal parietes (whether posterior, lateral, or anterior) muscular rigidity is the rule. Rigidity of the psoas should always be tested for by extending the right thigh with the patient on the left side. Rigidity of the quadratus lumborum should be present with an inflamed ascending appendix. It is difficult to ascertain but may be surmised if deep resistance is felt on pressing the fingers forward from beneath the lower posterior costal margin.

Fever. Fever may not be present at the beginning of the attack but frequently develops before twenty-four hours have passed. Before rupture has occurred the temperature is not usually much above normal, 2°F or 3°F being the average elevation. Mistakes are liable to be made due to the fact that the temperature is not elevated at the time of the onset of the pain. In any suspected case the temperature should be taken every two to four hours, and if it rises in a gradual manner it is a point in favor of appendicitis. If at the very beginning of any attack of acute abdominal pain the temperature is considerably raised (i.e., 103°F or 104°F.), or if a rigor has accompanied the onset of pain, the presumption is against appendicitis. Temperatures of this degree are not uncommon, however, once perforation has occurred, but it is to be kept in mind that perforation is extremely uncommon before twenty-four to thirty-six hours of the onset of symptoms.

Constipation. Though the patient frequently complains of constipation, especially during the early phase of visceral pain, numerous cases occur, particularly in children, in which an attack of diarrhea ushers in the attack.

Pulse. The pulse is only slightly, if at all, accelerated in the early stage; it may be normal in every way, even though the temperature is raised. *Any continued or decided acceleration of the pulse either corresponds with the occurrence of local peritonitis or indicates an appendix distended with infective material.*

Distention. When the appendix is acutely inflamed, *gaseous distention of the cecum* is sometimes present; this *local distention* is probably the result of a reflex adynamic ileus. It is more likely to be present when the appendix is retrocecal in position and closely embedded in the wall of the cecum. It gives rise to a local swelling with a tympanitic note on percussion. Diffuse and severe distention is encountered only in late cases of perforation.

Testicular symptoms. In the male, testicular symptoms are sometimes produced by an inflamed appendix even when unperforated. There may be pain in either the right or the left testicle, or in both, or the patient may say that the right testicle was retracted at a certain stage of the disease. It is likely that this is a pain referred from the appendix, since the tenth dorsal spinal segment apparently supplies both viscera. The direct stimulation of the genitocrural nerve by inflammatory exudate might account for testicular retraction.

THE ORDER OF OCCURRENCE OF THE SYMPTOMS

This is of the utmost importance in diagnosis. The march of events is:

1. Pain, usually epigastric or umbilical.
2. Anorexia, nausea, or vomiting.
3. Tenderness—somewhere in the abdomen or pelvis.
4. Fever.
5. Leukocytosis.

The symptoms occur almost always in the above order, and when that order varies, the diagnosis should be questioned. Moreover, leukocytosis is rightly put last in this sequence, for it probably means that local peritonitis has begun; *nowadays acute appendicitis should be diagnosed in many instances before there is demonstrative leukocytosis.* If fever precedes the onset of pain, or if vomiting accompanies or precedes the first bout of pain, it is generally not appendicitis with which we are dealing.

It is a fact worthy of remembrance that an acute attack of appendicitis often starts in the middle of the night and may awaken the patient out of sleep.

There are two or three facts about the *rectrocecal* appendix that need special mention. Pain is usually less, and is often from the first felt only locally. Vomiting is not so frequent, and generally the muscular rigidity over the diseased focus is less than would be expected for so advanced a lesion.

The diagnosis of appendicitis in the stage prior to perforation depends therefore upon certain constant and other inconstant features. Epigastric pain, nausea or vomiting, right iliac pain and low-grade fever, in that symptom sequence, are almost constant, and local tenderness, either on deep pressure in the right iliac region, in the flank, or by rectal examination, is invariable, except in the first hour or two of the attack. Local rigidity is common but not constant. The other symptoms mentioned above are inconstant, while the pulse rate is usually normal.

Although some have suggested that barium enema can be used to diagnose acute appendicitis by reason of failure to fill the appendix, most clinicians are loath to apply the large amounts of required pressure. In addition, failure to fill the obliterated lumen of an uninflamed appendix may confuse the issue. Occasionally, the presence of a calcified appendicolith on a plain film may be helpful if the clinical picture is consistent. An ultrasonographically demonstrable, thick-walled appendix may sometimes aid in diagnosis but is not an invariable finding in acute appendicitis.

In those rare cases in which, owing to a developmental anomaly, the cecum and appendix lie in the left instead of the right iliac fossa, the local symptoms and signs will be found on the left side. This condition is sometimes accompanied by dextrocardia, the existence of which should make one particularly careful in investigation of abdominal pain.

Diagnosis after perforation has occurred

A number of cases of acute appendicitis come to the surgeon after the appendix has perforated at the site of a patch of gangrene. One reason for this is that practitioners fail to realize that the symptoms of appendicitis as often described are those that accompany late appendicitis with perforation. Local abscess means perforation of the appendix, and some estimate may thus be formed of the relative proportion of perforated and unperforated cases. There are still too many cases that do not reach the surgeon until an abscess or peritonitis has developed. It must be allowed that in a few cases the first symptoms that the patient complains of seem to be those due to a perforation, but these cases are the exceptions.

The symptoms and course of illness consequent on appendicitis with perforation of the appendix are those already described, with the addition of the *symptoms due to local or diffuse peritonitis.* Perforation with the presence of a mass or with generalized peritonitis usually does not occur until at least forty-eight hours after the onset of symptoms. There are usually an accentuation of pain and renewal of vomiting when the perforation occurs, but the exact symptoms vary according to the position of the appendix and the nature of the protective peritoneal reaction.

Around the perforation itself a localized abscess may form, but sometimes a piece of omentum may seal the opening or, more rarely, the infection may spread quickly and widely without the formation of any or many adhesions. A definite lump is almost invariably indicative of a perforation. In the absence of a perforation, a lump rarely is caused by a thick and edematous cecum, or a cancer.

There are two main divisions of the pathological states consequent upon perforation, depending upon the position of the appendix itself:

1. When the appendix is above the brim of the true pelvis—the iliac appendix.
2. When the appendix lies wholly or partly in the true pelvis—the pelvic appendix.

THE ILIAC APPENDIX

Perforation of an appendix lying in the iliac fossa leads to rigidity of the overlying part of the muscular wall of the abdomen. When seen a little later, a lump may be felt and the rigidity may have lessened. Later still there may be simply a tender lump.

In addition, there will be fever (higher, as a rule, than before perforation) and certain localizing signs varying according to the position of the appendix (Fig. 17).

1. When the appendix perforates retrocecally there may be a mass that may be resonant on percussion, owing to the intervening cecum. One rarely perceives a mass sooner than seventy-two hours after the onset of symptoms, and sometimes not until the patient is anesthetized, because abdominal rigidity obscures the lump. The infection will cause inflammatory edema of the iliacus and quadratus lumborum and adjacent parts, and tenderness will be elicited on pressing the fingers forward below the right costal margin at the outer border of the erector spinae. A perforated retrocecal appendix is particularly likely to be missed in stout elderly patients in whom the condition may not come to the surgeon's attention before a definite swelling is palpable.

2. The appendix may lie in a position parallel with the cecum and ascending colon but lateral to them. The symptoms are then similar to those just described, except that rigidity of the lateral and anterior abdominal walls is more evident, and any lump is more easily felt because the cecum does not mask it.

3. The conditions resulting from perforation of an appendix lying in the iliac fossa on the iliacus or psoas are reasonably characteristic. Immediately after the perforation there will be intense rigidity of the abdominal wall over the right iliac region and great tenderness on pressure over the same area. (Very rarely, chiefly in patients

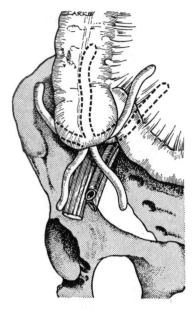

Fig. 17. The various possible positions of the appendix vermiformis.

suffering from severe toxemia, rigidity that had at first been present disappears on perforation of the appendix.) After a certain time, if there is suitable resistance to the infection, the peritoneal reaction becomes limited, and the rigidity usually diminishes somewhat, allowing the palpating hand to feel a tender lump—due either to a small local abscess or to a mass of omentum wrapped round the inflamed, perforated appendix. Obviously, if the infection is not walled off, generalized peritonitis rather than a mass may result.

There are two special symptoms that may help exact localization in this region. The irritation and reflex rigidity of the iliopsoas frequently cause the patient to hold the right thigh flexed, or with a lesser degree of irritation pain may be felt if the right thigh is fully extended as the patient lies on the left side. This sign is sometimes of great value.

In a few cases, irritation of the lateral femoral cutaneous nerve as it crosses the iliacus is evidenced by pain and hyperesthesia along the distribution of that nerve (the lateral thigh).

4. When the appendix lies so that the tip is directed medialward, the result of its perforation varies greatly according to whether it is behind or in front of the ileum. If *it is behind the ileum* localization of the inflammatory process usually results, but the swelling is not so readily felt since the ileum covers and masks it. But tenderness and rigidity may be present, and an indefinite lump may be felt, while the test for psoas irritation may help in diagnosis.

The ureter crosses the pelvic brim in close relationship to the medially directed appendix. Occasionally, pain on micturition may be produced, presumably by irritation of the ureter.

It is well to call attention to the lack of physical signs and the paucity of symptoms that may accompany an attack of acute gangrenous appendicitis when the appendix is fixed behind the end of the ileum. In the early stages of the attack the ileum is irritated and the bowel contents pass on quickly, giving rise to the passage of several small motions; in the later stages, after the appendix has perforated, the subsequent symptoms of obstruction of the lower ileum may mislead the observer as to the cause of the trouble.

It will be noted that this is another example of the extra difficulty and danger in diagnosis when the appendix is close to or against a nondemonstrative area of the peritoneum.

If the appendix perforates while lying *in front of the ileum* there is great danger of very extensive peritonitis, but if the infection becomes localized diagnosis is fairly easy, for the formation of pus close to the abdominal wall leads to local boardlike rigidity and exquisite tenderness over the affected area. Psoas irritation will be absent.

THE PELVIC APPENDIX

The early symptoms of an attack of appendicitis when the appendix is situated in the pelvis are similar to those that ensue when it is situated above the pelvic brim, except that rigidity of the right iliac region of the abdominal wall is seldom present in the early stages, and the pain is more frequently felt in both left and right iliac fossae. Pain is not so readily localized in the right iliac fossa but is often felt on deep pressure at the brim of the true pelvis, and the epigastric pain may dominate the scene for a longer time.

The perforated pelvic appendix is one of the most easily overlooked, and therefore one of the most dangerous, conditions that may occur in the abdomen. The reason appears to be as follows. While the appendix is unruptured and tense, the pain due to the distention and peristaltic contraction is definite and severe, and is felt chiefly in the epigastrium or the umbilical zone. When rupture occurs, the epigastric pain diminishes and local pelvic peritonitis results on the right side of the pelvis or at the bottom of the pelvic pouch of peritoneum. *This is usually unaccompanied by rigidity of the lower abdominal muscles,* and since the pain of appendicular distention has ceased and the pain due to pelvic peritonitis at this stage is frequently very insignificant, the patient may seem better, and the examination of the abdomen may give little indication of the trouble in the pelvis. Sooner or later—usually within three or four days—either the peritonitis becomes definitely localized into a pelvic abscess of considerable dimensions or the inflammation may track upward toward the general abdominal cavity and give rise to generalized peritonitis with increasing pain, distention, and rigidity of the abdominal wall. If, because the patient came late for advice or sufficient attention was not paid to the preliminary symptoms, the prerupture stage of the inflamed pelvic appendix is missed, it is essential to diagnose the ruptured appendix as soon as possible after rupture before peritonitis has extended too far upward into the abdominal cavity. For this purpose it is important to pay attention to the anatomical position of the pelvic appendix, which lies in relationship with one or more of the following—*the pelvic wall, the rectum, and the bladder.* Irritation of the bladder or rectum may be signified by frequency of or pain during micturition, or by diarrhea or tenesmus, respectively. But more important is the fact that usually a tender swelling can be felt against the right pelvic wall by the finger inserted into the rectum. Moreover, when the ruptured appendix is adherent to the fascia covering the obturator

internus and the subjacent fibers of the muscle are affected by the
inflammatory edema, rotation of the flexed thigh so as to put the
muscle through its extreme movements (especially internal rotation) will cause hypogastric pain. In performing this maneuver it is
essential that the thigh be flexed so as to relax the psoas muscle.

By a careful consideration of the history and of the points just
mentioned, it should be difficult to miss the early inflamed pelvic
appendix. The appendix that is most likely to give rise to doubt in
diagnosis is one situated high up in the right posterior quadrant of
the pelvis, for in this situation there may be no localizing signs and
it may be difficult to feel the viscus per rectum.

The later symptoms resulting from rupture of a pelvic appendix
are either those of a large pelvic abscess or those of advanced pelvic
and hypogastric peritonitis, with increasing toxemia. Tenderness
and rigidity of the whole lower abdominal wall, distention, vomiting, and increased pain all give a clear picture of peritonitis.

It is noteworthy that as the peritonitic inflammation spreads
upward from the pelvis, it does so frequently on the left side first.
This is probably because as the pelvis fills with pus, the anatomical
path of least resistance is by the side of the sigmoid colon.

When a pelvic abscess has formed there are usually all the
symptoms of suppuration—fever, anorexia, and leukocytosis, while
locally there are tenderness and slight distention in the hypogastrium. Rigidity of the lower abdominal wall is quite frequently
absent, even when a pelvic abscess is present. Rarely, the temperature may be normal even in the presence of an abscess. An
ultrasound scan is *occasionally* of value in localizing an abscess in
the pelvis, whereas CT is usually more definitive.

Per rectum the bulge of the abscess can usually be detected, and
pain is produced by pressing on the bulging mass.

In women the intervention of the uterus makes bladder symptoms less likely to supervene in cases of appendicular suppuration.
The normal menstrual flow may be increased by the pelvic congestion, and the menstrual period may be precipitated thereby.

There is a rare condition of intestinal obstruction that may be
produced by pelvic appendicular suppuration at a late stage. Either

or both the sigmoid colon and the small intestine may be obstructed within the pelvis. This is most likely encountered in the aged or debilitated patient in whom the early stages of the disease were not detected. But if the inflamed pelvic appendix were diagnosed in the early stage, there would never arise any such dangerous late complication. In early diagnosis and operation lies the prevention of such conditions.

RECURRENT ACUTE APPENDICITIS

While there is agreement that chronic appendicitis will not cause prolonged abdominal pain lasting for weeks or months, it is clear that patients occasionally experience recurrent attacks of acute appendicitis from which they recover. The individual attacks do not differ from those of classical acute appendicitis. Because of the recurrent nature of the attacks, barium enema examinations are often performed on such patients, and sometimes demonstrate a characteristic extrinsic compression of the medial aspect of the cecum caused by a small abscess. Such patients are often thought to have regional enteritis but are cured completely by appendectomy.

7. The differential diagnosis of appendicitis

Diagnosis of appendicitis is usually not especially difficult. A considerable number of the acute abdominal emergency patients admitted to a hospital have appendicular inflammation. So frequent is this condition that it would almost appear that some do not trouble to attempt a differential diagnosis, a serious mistake.

The typical case with epigastric pain, followed by vomiting, succeeded by localizing of the pain in the right iliac fossa where tenderness can always, and rigidity of the overlying rectus can often, be made out, is sufficiently characteristic, even without the presence of slight fever, to make the diagnosis very probable. But there are difficulties that have to be discussed. *Acute appendicitis can mimic virtually any intra-abdominal process,* and therefore to know acute appendicitis is to know well the diagnosis of acute abdominal pain.

SYSTEMIC CONDITIONS OR INFECTIOUS CONDITIONS
THAT SIMULATE APPENDICITIS

Influenza. It is important first to make quite sure that one is dealing with a primarily abdominal condition. In the course of a definite attack of influenza abdominal pain may ensue, and during an outbreak of the disease there is grave danger that occasionally an attack of appendicitis may be overlooked and attributed to "abdominal influenza." But seldom in influenza is the abdomen alone attacked. Pain and tenderness in the right iliac fossa are sometimes present in influenza, though the abdominal pain is more likely to be general. Backache and pain in the eyeballs are more likely to be felt in an attack of influenza, and vomiting may precede the abdominal pain—a sequence seldom seen in appendicitis.

Diaphragmatic pleurisy. One must always exclude diaphragmatic pleurisy as a result of early basal pneumonia or pulmonary infarction before diagnosing appendicitis. Pain, tenderness, and muscular rigidity may all be noted in the right iliac region in thoracic disease, but sometimes firm continued pressure will enable one to feel deep into the iliac fossa without causing any increase of the pain. In cases of appendicitis, pressure over the left iliac fossa applied by fingers pressed deeply in and directed toward the right side will sometimes cause pain in the appendicular region—a sign that is absent in cases of pleurisy or pneumonia. In thoracic disease the respiration rate is usually increased, and more importantly, the abdominal wall moves freely with respiration. Of course, a careful examination of the chest, together with a chest X-ray if necessary, is *the* method of discrimination.

Spinal disease. Very rarely, spinal disease may cause pain referred to the appendicular region. An examination of the spinal column can easily and quickly be made and soon determines any lesion.

Typhoid. Typhoid fever, still prevalent in some parts of the world, is another general disease that is occasionally mistaken for appendicitis because of the abdominal pain and tenderness that are sometimes localized in the right iliac fossa. In most cases, however, there are general symptoms that enable the diagnosis to be made. Headache, general malaise, enlargement of the spleen, and the presence of a roseola should make one suspect typhoid, and the absence of a leukocytosis would exclude appendicitis. In children in whom the general symptoms of typhoid fever are often slight, and occasionally in adults attacked by the ambulatory type of the disease, mistakes might be made. If the symptoms have been present for a week when the patient is first seen, the agglutination reaction may be positive in respect to *Salmonella typhi* or *S. paratyphi A* or *B*. A blood culture of a typhoid patient might prove positive if taken within the first week of the disease. One must utter a warning against treating any doubtful case as typhoid without making a rectal examination. The irregular fever, tympanitic abdomen, and vague hypogastric pains that accompany a pelvic abscess may be, and have been, mistaken for typhoid fever, but a finger inserted into the anal canal will serve to distinguish the condition unless the abscess is due to leakage of a typhoid ulcer.

Other more common types of salmonellosis or shigellosis that are now generally acquired from contaminated food may produce nausea, vomiting, abdominal pain, and diarrhea. The last is usually profound, and the temperature is often 102°F to 103°F from the onset, serving to distinguish this condition from acute appendicitis.

Acute porphyria. In certain predisposed persons and under special circumstances the hemoglobin of the red blood cells may be broken down, with the formation of an abnormal amount of porpho-bilinogen and the ultimate excretion of certain porphyrins in the urine that, on standing, turn a dark brown or reddish brown color. These abnormal biochemical changes may be accompanied by alarming abdominal or neurological symptoms, particularly by severe attacks of acute intestinal colic; this has sometimes led to the erroneous diagnosis of acute appendicitis. Only the hepatic types of acute intermittent or cutanea tarda hereditaria porphyria and hereditary coproporphyria are associated with attacks of abdominal pain.

Acute porphyria occurs chiefly in those who have inherited a predisposition to it, and symptoms more commonly arise in the third and fourth decades of life. It has become more common during the last twenty years, during which there has been a greatly increased use of certain drugs, particularly the barbiturates and the sulfonamides, which have a tendency to precipitate attacks. The attack often follows drug taking or the administration of an anesthetic. The taking of alcohol may also cause an attack. Acute porphyria is more common in women and is especially common in South Africa.

The abdominal symptoms consist of severe colicky pain, vomiting, and constipation. The pain is central but may radiate widely. It has been mistaken for biliary colic. As a rule there is no distention, tenderness, or muscular rigidity, but the intensity of the pain, felt around the umbilicus, in the back, and sometimes in the limbs, may be alarming. In severe cases there may be muscular weakness, and mental symptoms, such as hallucinations, may even suggest hysteria. Slight fever and jaundice may occur, and there may be a leukocytosis.

Even before the urine turns brown, the porphobilinogen may be detected in urine by chemical means. If porphobilinogen is present, it is specific for acute intermittent porphyria and for the cutanea tarda hereditary form during an acute attack. In patients with hereditary coproporphyria, there is an increased excretion of coproporphyrin III in the stool and the urine, which can be found by appropriate testing.

Acute porphyria may stimulate acute appendicitis or even intestinal obstruction, but the pain does not localize in the right iliac fossa, and as a rule there is no distention. Careful inquiries should be made as to any similar attacks in the patient or in any of her relatives.

MINOR ABDOMINAL DISTURBANCES

In the early stages of appendicitis, before there is any or much congestion or inflammation of the peritoneum, and when muscular rigidity is usually absent, appendicitis is often mistakenly diagnosed as "gastroenteritis" or "gastritis."

It is not facetious or cynical to say that the diagnosis of gastritis or gastroenteritis is usually made in the emergency ward by a young physician who is "not impressed" by a patient's abdominal pain or physical findings. The failure to impress reflects the erroneous idea that an "acute abdomen" (or one requiring a surgeon) is so catastrophic that the patient must complain of severe pain and have boardlike rigidity of the abdomen. In my experience, the diagnosis of gastroenteritis in the emergency ward is so often incorrect as to raise a serious question whenever the emergency physician comes to this conclusion. The diagnosis is usually made in a patient with abdominal pain, nausea, vomiting, diarrhea, or any combination of these that cannot be attributed to a more definite condition by the examining physician. My teaching for many years has been that the emergency room physician should be uncomfortable with the diagnosis of gastritis or gastroenteritis.

Acute gastritis, which may be caused by bacteria such as *Helicobacter*, viruses, excessive alcohol intake, or a variety of drugs, is usually marked by intense nausea and vomiting. While a dull, gnawing epigastric discomfort is often present, it is not localized and does not become so, nor are abnormalities present on careful examination of the abdomen.

Viral or bacterial (*Campylobacter, Shigella, Salmonella*) infections of the more distal portions of the gastrointestinal tract are more frequently associated with fairly profound diarrhea. Tenderness and rigidity are usually minimal or absent. If a smear of the loose stool shows sheets of leukocytes, the conclusion that enteritis is present is on firmer ground than when gastroenteritis becomes an exclusionary diagnosis.

As in adults, the diagnosis of gastroenteritis in children should not be made with peace of mind. Any child previously in good health who is suddenly taken with abdominal pain and loss of appetite, has nausea or vomits, and at the same time shows definite deep tenderness in the right iliac fossa, even if the pulse is normal and the temperature is not elevated, is most probably suffering from appendicitis.

Infective hepatitis. It is well to remember also that some patients who suffer from infective hepatitis complain of abdominal pain with anorexia and nausea in the early stage of the disease, and it is easy for this to give rise to a suspicion that the appendix may be diseased. As a rule, however, the pain does not localize in the right iliac fossa. Inquiry may be made as to the prevalence of other cases of hepatitis, and, of course, when the jaundice appears the problem will be solved.

Diabetes. In an acute crisis of diabetic ketosis there may be severe abdominal pain and sometimes rigidity of the abdominal wall simulating a local inflammatory lesion such as appendicitis. The finding of sugar and acetone in the urine will at once put the observer on guard. If the abdominal symptoms do not rapidly subside under treatment for the ketoacidosis, further examination should elicit the true cause of the condition, for a diabetic patient *may* develop intra-abdominal lesions. The latter may indeed pre-cipitate the ketoacidosis.

COMMON ABDOMINAL OR RETROPERITONEAL CONDITIONS
THAT MAY SIMULATE APPENDICITIS

When the local signs and symptoms of appendicitis are well devel-oped (pain, tenderness, hyperesthesia, rigidity), there are many conditions that have to be excluded and for which it may be mistaken. When the local signs are very clear, the appendix is usually either perforated or in danger of perforating. To catalog the diseases that *may* simulate or be simulated by appendicitis is to enumerate all the chief acute abdominal diseases. This is obviously of little practical value. It will be better, therefore, to give only the more common conditions causing mistakes and to group them according to the position of the appendix:

The ascending appendix (retrocecal or paracecal):

Cholecystitis
Inflamed duodenal ulcer
Perforated gallbladder
Perinephric abscess
Hydronephrosis

Pyonephrosis
Pyelitis
Stone in the kidney
Torsion of the omentum

Iliac position of appendix:

Leaking duodenal ulcer
Crohn's disease
Cecal or ileocecal carcinoma
Tuberculous ileocecal glands
Stone in ureter
Yersinia infections

Inflamed Meckel's diverticulum
Psoas abscess
Tuberculous hip disease
Rupture of rectus muscle
Simple ulcer of the cecum

Pelvic position of appendix:

Intestinal obstruction
Diverticulitis with abscess

Perforation of a typhoid ulcer
Gastroenteritis

In women:

Ectopic gestation
Twisted pedicle of an ovarian cyst or
   of a hydrosalpinx
Rupture of a pyosalpinx

Salpingitis
Ruptured follicular cyst (*Mittelschmerz*)

Ruptured corpus luteum cyst

Before diagnosing appendicitis in tropical climes one would also have to exclude:

Amebic typhlitis
Hepatitis

Leaking liver abscess
Malaria

When the local manifestations have spread widely and the patient first comes under observation with generalized peritonitis, it is necessary to distinguish the condition from all the various causes that may lead to such a pathological picture.

Late cases with extensive peritonitis must be distinguished from:

1. Acute intestinal obstruction (Chapter 13).
2. Thrombosis or embolism of mesenteric vessels (pp. 166, 201).

3. Acute pancreatitis (Chapter 9).
4. Pneumococcal peritonitis (p. 243).
5. Pylephlebitis (p. 236).
6. General peritonitis (ruptured gastric, duodenal, typhoid ulcers, etc.; pp. 239–242).

Differential diagnosis of the perforating or perforated ascending appendix

The gallbladder, the duodenum, and the kidney are the viscera in anatomical proximity to the ascending appendix, and inflammation of them or their surroundings may cause difficulty in diagnosis.

Cholecystitis. Cholecystitis may very closely simulate appendicitis. Pain, vomiting, fever, constipation, and local tenderness on the right side of the abdomen are present in both cases. An enlarged, inflamed gallbladder frequently comes down into the right lumbar region but more usually enlarges in the direction of the umbilicus. In thin patients without rigidity of the abdominal wall diagnosis is usually easy, for the tender, rounded gallbladder may be felt continuous with the liver and perhaps moving with respiration. The pain of cholecystitis is usually a little higher than that of an ascending appendicitis, and there may be pain of a segmental nature referred to the right subscapular region, especially if a stone is impacted in the cystic duct. There may be resonance of the ascending colon over an inflamed retrocecal appendix. There is never resonance in front of an inflamed gallbladder, which is usually on a plane anterior to the cecum, colon, and appendix. In very stout patients and in those with very rigid abdominal muscles, it may on occasion be almost impossible to determine whether the appendix or gallbladder is at fault without giving an anesthetic, unless the previous history is clearly indicative of one or the other condition.

When with cholecystitis a stone is simultaneously impacted in the cystic duct, the constant pain accompanied by retching, with deep tenderness in the right hypochondrium and right subscapular region, are sufficiently diagnostic and clearly differentiate the condition from appendicitis.

Inflamed duodenal ulcer. Periduodenitis around an inflamed duodenal ulcer should be distinguishable by the characteristic history elicited by careful questioning. The pain of duodenal ulcer comes two or three hours after taking food and is relieved by taking food.

Pyelitis. Acute right-sided pyelitis is frequently mistaken for appendicitis, and occasionally operations are unwisely undertaken because insufficient attention is paid to the symptoms. The points in differential diagnosis can be tabulated as follows:

Acute pyelitis	*Appendicitis*
Initial rigor common	Rigor unusual
Temperature of 103°F or more	Temperature as high as 103°F
Pain on micturition	uncommon
Increased frequency of urination	Urinary symptoms inconstant
Abdominal muscles often lax	Local rigidity frequent
Pus or bacteria in urine	Usually no pus or bacteria in urine

The reader will note that it is unusual for a rigor to occur in a case of appendicitis. If a rigor or rigors do occur in a patient who is undoubtedly suffering from appendicitis, then it is possible, nay likely, that there is an immediate danger of portal pyemia. At the operation, special attention should be paid to the condition of the tributaries of the superior mesentric vein, and during convalescence the possibility of hepatic abscess should be borne in mind.

The symptoms of acute pyelitis may be produced by the presence of bacilli in the urine without formation of pus. In such cases there is usually turbidity or opalescence of the urine, which may acutely have developed an offensive odor. One must not forget also that an inflamed appendix lying in front of the renal pelvis may actually cause an acute pyuria without a bacilluria. *If the urine is carefully examined as a routine, there will seldom be any difficulty in diagnosis.*

Hydronephrosis. Acute right-sided hydronephrosis is sometimes misdiagnosed as appendicitis with abscess formation. A hydronephrosis forms a rounded, tense, tender swelling that occupies the lateral aspect of the abdomen and can be felt well back in the loin. The swelling is sometimes freely movable and usually round in shape. It may be possible to feel a depression (corresponding with the hilum) on the medial side. The pain is sometimes of the type of renal colic and *there are usually urinary symptoms*—scanty urine, pain during or frequency of micturition, etc. It may be possible to ascertain a history of previous attacks. Rigidity of the abdominal wall over the swelling is usually absent.

Pyonephrosis. Acute pyonephrosis forms a swelling similar to a hydronephrosis, but it is usually more tender and more fixed, and the general signs of constitutional disturbance are much greater; for example, there are high fever and other symptoms of toxemia. There may be pus in the urine if the ureter is not completely obstructed.

Stone in the kidney or ureter. In those unusual cases in which appendicitis is accompanied by pain in the right testis, it may closely simulate renal colic. An X-ray film may show the stone and differentiate. In about 20 percent of cases, small ureteric calculi do not show on an X-ray film, and an intravenous pyelogram or ultrasound scan of the kidney must then be done. In one patient under my care, acute pain in the right loin radiating to the right testis, accompanied by fever and some muscular rigidity, caused me to diagnose renal colic. A radiograph showed a large shadow a little external to the normal line of the ureter. Operation revealed a normal kidney and ureter, but a very inflamed appendix with a

large, calcareous fecolith in the mesoappendix. Fortunately such
cases are rare, although recently it has been recognized that the
presence of a characteristic laminated fecolith in the region of the
right iliac fossa in a patient with a compatible history usually helps
clinch the diagnosis of acute appendicitis.

Perinephric abscess. A suppurating retrocecal appendix may form
an abscess in the neighborhood of the kidney and may be difficult to
distinguish from a perinephric abscess of metastatic origin. But the
latter is insidious in origin, while appendicitis usually has an acute
history of onset. In both cases there will be pain on pressing forward
in the erector-costal angle below the last rib. Some patients with an
inflamed retrocecal appendix present atypical symptoms, have no
initial epigastric pain, do not vomit, and present no rigidity over the
inflamed area. *These cases, however, are much more rapid in
development than the usual metastatic perinephric abscess.* A small
retrocecal or retroileal abscess of appendicular origin may be easily
be overlooked (Fig. 18).

Torsion of the omentum. Torsion and strangulation of the whole or
of a portion of the omentum may simulate appendicitis. The part
affected is usually to the right of the midline, and pain and
tenderness will be noted to the right of the umbilicus. Vomiting is
less common than in appendicitis, but differential diagnosis before
operation may be impossible. The pain due to torsion of the
omentum is usually relieved when the patient lies down. Sometimes there may be a history of undue exertion, and there may be a
doughy tumor felt in the midline or on the right side of the
abdomen.

Differential diagnosis of the inflamed iliac appendix

Inflammation of the iliac appendix is the most easy condition to
diagnose, though there are many pitfalls.

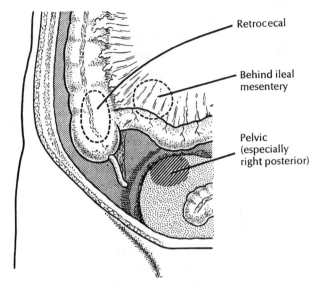

Fig. 18. Those sites where an abscess resulting from appendicitis may sometimes be overlooked (see pp. 82 and 102).

Perforated duodenal ulcer. A perforated duodenal ulcer is frequently misdiagnosed as appendicitis. The escaping contents travel down to the right iliac fossa and give rise to all the signs of inflammation of the appendix. It may be possible to obtain a typical duodenal or appendicular history. The initial pain at onset is greater in the duodenal condition, and there will also be definite right hypochondriac tenderness. Pain felt on the top of the right shoulder would be more in favor of a perforated duodenal ulcer. If there is any obliteration of liver dullness in the absence of general abdominal distention, a duodenal (or gastric) perforation is almost certain. The presence of free air on X-ray is virtually pathognomonic because free air almost never is encountered in acute appendicitis, even with perforation.

Crohn's disease. Crohn's disease (regional enteritis) may closely simulate acute appendicitis and appears to have become more frequent during the past few years. There may be abdominal pain, vomiting, and local tenderness in the right iliac fossa, and sometimes a palpable lump may be felt. Unless the patient has been previously under treatment, the condition may be impossible to differentiate before the abdomen is opened. The key to differential diagnosis is the almost invariable history of previous attacks. The presence of a mass, especially if it is known to have been present before the attack, points strongly to Crohn's disease. If the condition is subacute, then the history of attacks of colicky pain, the occasional bouts of diarrhea, and (if a barium meal has been given) the very characteristic radiographic picture of a narrowed end of the ileum will reveal the true state of affairs. As mentioned before, distinguishing between recurrent attacks of acute appendicitis and Crohn's disease may be impossible. Barium studies should differentiate these conditions, however. The correct diagnosis has, however, frequently been made only after opening the abdomen for a supposed attack of appendicitis. In recent years, acute ileitis caused by *Yersinia* organisms has been differentiated from Crohn's disease and may account for a large number of the cases previously called *acute* Crohn's disease. Differentiation from acute appendicitis prior to operation is usually impossible, but cultures of ileal lymph nodes or complement fixation tests will disclose the true state of affairs. Most cases of *acute* Crohn's disease do not become chronic and therefore, in my opinion, most of them are likely to have been instances of *Yersinia* or some other infection.

Carcinoma. Carcinoma of the cecum may cause subacute attacks of severe pain and local right iliac tenderness, but the pain is more griping and persistent than in any but the most acute attacks of appendicitis. Cecal carcinoma occasionally causes acute appendicitis by occluding the orifice of the appendix. Carcinoma of the cecum or of the ascending colon may become adherent to the anterior abdominal wall and may closely simulate an appendicular abscess. The age of the patient (usually over fifty), previous attacks suggestive of obstruction, and noticeable loss of weight may put one on guard, and the history of progressive anemia and of attacks of diarrhea would favor cancer of the cecum.

Ileocecal tuberculosis of the nonobstructive type may cause symptoms similar to those of carcinoma.

Nonspecific adenitis of the ileocecal glands may simulate the earliest stage of appendicitis by giving rise to colicky abdominal pain and local tenderness in the right iliac region, but the condition does not progress and there should not be vomiting or local resistance. Again, many of these cases are now known to have represented instances of *Yersinia* infection.

Psoas abscess and *tuberculous hip disease* may cause irritation of the iliopsoas, with flexion or limitation of extension of the right thigh and tenderness, resistance, and fullness in the right iliac fossa, but the general onset and subacute nature of the illness, together with careful examination of the spine and hip, in most cases easily establishes the diagnosis. A radiograph should be taken if doubt exists. These conditions are now becoming rarer owing to the more effective treatment of tuberculosis.

Stone in the ureter. As already mentioned, a stone passing down the ureter causes pain, not always typical of renal colic, referred approximately to that section of the ureter in which the stone is lodged. Quite frequently this causes a diagnosis of appendicitis to be made. Urinary symptoms (frequency, pain, hematuria), pain in the testicle, absence of rigidity over the painful area, and a previous history of attacks suggestive of renal colic should put one onto the right line of diagnosis; and if expert assistance is available, radiography may serve to demonstrate a calculus. Fever is unlikely to be present in the case of ureteral stone.

Meckel's diverticulum. Inflammation of a Meckel's diverticulum gives rise to symptoms that may be almost indistinguishable from those of acute appendicitis. Since the diverticulum is usually nearer the center of the abdomen and is not usually adherent anywhere, the pain may be central and poorly localized for a longer period. It should be emphasized that Meckel's diverticulitis is not common.

Rupture or hematoma of the right rectus muscle. Rupture of the lower segment of the right rectus muscle may also lead to local signs similar to those caused by appendicitis, but the history of the onset should serve to distinguish the conditions. Rupture of the rectus is prone to follow a great or sudden muscular effort or may be due to a severe bout of coughing. Vomiting and intestinal symptoms will be absent. Severe bleeding from the torn deep epigastric artery may cause shock; a CT scan will show a gap in the rectus muscle as well as the hematoma. In patients taking anticoagulants, hematoma may occur either spontaneously or with only the slightest exertion. The signs and symptoms are identical to those of rupture of the muscle.

Simple ulcer of the cecum. These lesions may cause inflammation or perforation that is indistinguishable from acute appendicitis.

Differential diagnosis of the pelvic appendix

DIFFERENTIAL DIAGNOSIS IN THE MALE

Inflammation of an appendix situated in the pelvis gives rise to many mistakes in diagnosis. In women there is some excuse for this, but in men there are comparatively few conditions that cause severe acute pelvic pain, and mistakes should not occur so frequently. The chief conditions for which it may be mistaken are:

1. Obstruction of the large bowel (carcinoma, volvulus).
2. Obstruction of the ileum.
3. Diverticulitis.
4. Stone in the lower part of the ureter.
5. Gastroenteritis.
6. Pelvic abscess.

Obstruction of the large bowel. Obstruction of the large bowel, causing hypogastric symptoms, is commonly due to carcinoma of the sigmoid or rectum or to volvulus. The onset of both of these conditions is usually preceded by a time of subacute obstruction with attacks of abdominal pain and distention, and in both cases *distention is an early feature of the acute attack*. In pelvic appendicitis the system sequence is fairly constant, and *distention is not an early symptom*. In both cases rectal examination will reveal pelvic tenderness, but in obstruction there may be greater ballooning of the upper part of the rectum, while in appendicitis there is often a tender lump in the right side of the pelvis, and the thigh-rotation test may be positive. In appendicitis also there is not complete obstruction. Fever is usually absent in obstruction and present to some extent in appendicitis.

Obstruction of the ileum. Obstruction of the ileum, accompanied by tenderness in the hypogastrium, is frequently due to adhesions caused by previous abdominal operations of any kind. Distinction is made chiefly by noting that in obstruction there is greater acuteness of pain, which is of a spasmodic nature, and by observing the frequency and character of the vomit, which in obstruction gradually becomes yellowish and finally feculent—a change that never happens in appendicitis until extensive peritonitis has developed. In intestinal obstruction the pain is seldom localized in the right iliac fossa, as in appendicitis, but after distention has supervened diagnosis is made much more difficult. In small-bowel obstruction also the temperature is usually normal at onset and does not frequently become febrile as is usual in appendicitis. Frequency of micturition or pain during the act may occur in appendicitis owing to irritation of the bladder. Sometimes also with a retroileal situation of an inflamed appendix there may ensue an obstruction of the lower ileum, especially in the aged, in whom the early diagnosis of acute appendicitis often goes unnoticed or undiagnosed.

Diverticulitis. Diverticulitis of the pelvic colon may cause either obstructive or inflammatory symptoms. When causing obstruction it closely resembles carcinoma, but when local inflammation and abscess result, the symptoms and signs are very similar to those of pelvic appendicitis.

In fact, it is well to regard sigmoid diverticulitis as *left-sided acute appendicitis.* Pathogenetically, the two diseases often arise from the same mechanism, namely, obstruction of a diverticulum of the colon. Symptomatically, the sequence of symptoms so characteristic of acute appendicitis is usually observed in diverticulitis, but with a few modifications worthy of note: (1) the early visceral pain is of the same character but is more likely to be hypogastric in sigmoid diverticulitis and epigastric in appendicitis; (2) anorexia, nausea, and vomiting usually follow the initial pain but are somewhat less prominent than in appendicitis; (3) as in appendicitis, there is usually a shift of pain and tenderness, but in this case to the left iliac fossa or left suprapubic area, although on occasion only pelvic tenderness is evident; (4) a more pronounced change in bowel habit, most often diarrhea, occurs in diverticulitis; (5) fever and leukocytosis are more pronounced in most cases of diverticulitis; (6) and finally, a mass in the left iliac fossa or in the cul de sac of Douglas is a common finding in diverticulitis. It should, of course, be emphasized that diverticulitis is chiefly encountered in older persons, although we have made the diagnosis in a twenty-four-year-old man and have cared for a ninety-year-old patient with acute appendicitis. Previous attacks of diverticulitis are often reported but may occur many years apart. When the sigmoid colon is redundant and lies well to the right, or when the appendix points well to the left, the confusion in diagnosis can be considerable.

It has recently been pointed out than an "attack" of diverticulitis (i.e., an episode), which can be diagnosed clinically, is almost always the result of a perforation of the diverticulum with the development of a well-localized peridiverticular abscess. This explains the greater likelihood of fever, leukocytosis, diarrhea, and the presence of a mass. On occasion the peridiverticular abscess may rupture and produce peritonitis without a clear opening in the colon, probably because the diverticulum is usually obstructed by a fecolith. Even more rarely, there is a free rupture of the diverticulum itself into the peritoneal cavity. Differentiation of these types of rupture is prognostically important since mortality due to the less common free perforation is so much greater.

The differential diagnosis between diverticulitis and appendicitis may be so difficult, and the treatment so different, that there are cases in which a gentle enema done with a water-soluble radio-opaque material will solve the problem by demonstrating a tiny extracolonic tract arising from the pelvic colon. Only those cases in which the inflammatory process is *very well localized* should be considered for a radiopaque enema and this decision should be made only by a surgeon. A CT scan with contrast may be very useful in helping to make the distinction.

Perforation of a typhoid ulcer. When, in a patient suffering from a mild attack of typhoid fever a subacute perforation occurs in the lower ileum, a pelvic abscess may result that may at the time of examination be indistinguishable clinically from that due to a perforated appendix. But the diagnosis will be helped by the previous history.

Stone in the lower part of the ureter (see p. 98). When a stone is near the bladder there are often additional symptoms, frequency, strangury, pain in the penis, or emissions, which may point to the genitourinary system.

Gastroenteritis. When the inflamed appendix is in the pelvis and lies against the rectum, "diarrhea" in the form of frequent loose, mucoid movements is common, and for that reason the diagnosis of gastroenteritis is frequently made. True pelvic tenderness and/or a lump in the cul de sac are never present in gastroenteritis and should serve to distinguish between the two conditions.

Pelvic abscess. Instances occur in which the acute symptoms due to appendicitis are transient, the inflammation becomes limited by adhesions to contiguous viscera, and perforation may occur without the supervention of extending peritonitis. In such circumstances the patient may not seek medical advice until an abscess has developed. When the abscess is retrocecal, the observer will usually detect a lump in the right iliac fossa. When the abscess has resulted from the perforation of an appendix situated in the pelvis, the resulting abscess may reach large dimensions without being palpable above the pubis; as the abscess increases in size, it may cause pressure upon the rectum and the lower coils of the ileum, leading to a true intestinal obstruction. The course of the whole illness may be so insidious that the patient may first present symptoms of obstruction of the ileum—colicky pain, a ladder pattern of distention, etc. The importance of a rectal examination is obvious, and sometimes an X-ray examination will be of assistance.

Such a pelvic abscess may attain a large size without giving rise to obvious symptoms until its pressure on the end of the ileum causes definite obstruction. The cause of the alarming symptoms may not be obvious until a careful rectal examination has been made. A boy was once admitted to my wards with symptoms of acute intestinal obstruction. No apparent cause was discovered until a rectal examination revealed a large pelvic abscess. Some days previously, the patient had symptoms that were in keeping with an acute attack of appendicitis that had subsided before the obstruction had developed. The abscess was opened per rectum, the distention and vomiting subsided, and the obstruction was relieved without any need for an abdominal incision.

DIFFERENTIAL DIAGNOSIS IN THE FEMALE

The female reproductive organs add considerably to the difficulty of diagnosis of pelvic appendicitis. Acute pain referable especially to the hypogastrium and pelvis may be due to:

1. Uterine colic (dysmenorrhea or threatening abortion).
2. Twisted pedicle, inflammation, rupture of an ovarian cyst, or torsion of a normal or slightly enlarged ovary.
3. Ectopic gestation.
4. Twisted or inflamed fibroid.
5. Twisted hydrosalpinx.
6. Salpingitis or pyosalpinx.
7. Rupture of an endometrioma.

Dysmenorrhea. Dysmenorrhea, with its periodicity, lack of signs on local examination, and pain referred to the lower lumbar and sacral regions as well as the hypogastrium, should not cause serious difficulty in diagnosis.

Threatened abortion. In threatened abortion, the previous amenorrhea, bleeding, character of the pain, and absence of local signs serve to distinguish the condition. A positive pregnancy test will be helpful.

Ectopic gestation. Ectopic gestation is frequently misdiagnosed as appendicitis, but there is almost invariably some menstrual irregularity, often a history of a fainting attack, and generally some anemia, and the symptom sequence of appendicitis is not usually ascertainable. The uterus may be displaced by the extravasated blood. In unruptured cases the enlarged tube may be felt as an abnormal mobile and tender swelling to one side of the uterus (see Chapter 18), and the pregnancy test is positive.

Ovarian cyst and hydrosalpinx. In the case of an ovarian cyst or hydrosalpinx with a twisted pedicle, diagnosis is made chiefly by the fact that with the twisted viscus *the pain and vomiting come on simultaneously* (or almost so), so that the proper appendix symptom sequence is wanting; moreover the vomiting or retching is usually more frequent and more persistent than in appendicitis. In the case of an ovarian cyst it may have been previously known that there was a tumor, and a definite tender swelling may be made out from the time of onset of the symptoms. This swelling may be situated in the midhypogastrium or to one or the other side or may be limited to the pelvis. Torsion of a normal ovary may occur but is rare. If this happens in early pregnancy, there may be difficulty in distinguishing it from an early ectopic pregnancy or a pelvic appendicitis.

The symptoms of a ruptured follicular cyst (*Mittelschmerz*) or a ruptured corpus luteum cyst may be difficult to distinguish from acute appendicitis if on the right side. *The timing as related to the menstrual cycle is of the utmost importance.* In *Mittelschmerz* the pain occurs almost precisely at midcycle, whereas in rupture of a corpus luteum cyst the pain develops around the time of the menses. Nausea and vomiting are less frequent than in appendicitis, but the patient may show evidence of some blood loss. The usual appendix symptom sequence is absent.

With a twisted fibroid the symptoms are not actually so acute, and the presence of the fibroid will perhaps have been known previously. It may be impossible to differentiate between a twisted fibroid and an ovarian cyst with twisted pedicle.

Acute salpingitis. Acute salpingitis is one of the most difficult conditions to distinguish from appendicitis. Occasionally it is difficult to say in which of the contiguous organs the inflammation started. Distinction may usually be made by considering the following points. Acute salpingitis does not so frequently cause epigastric pain at the onset, and vomiting is less frequent. The salpingitic pain is frequently felt on both sides from the onset, and there may be greater tenderness in the left iliac region than the right. The history is often unreliable, but the presence of a vaginal discharge is a valuable guide, and examination should be directed to this point. It has been stated that with salpingitis pain is more often felt down the thigh even as far as the knee, but this is certainly not a constant symptom. If in doubt, it is better that operation should be undertaken, but if there is no reasonable doubt that the condition is acute salpingitis most surgeons consider nonoperative treatment preferable. It has been said that an elevation of the erythrocyte sedimentation rate (ESR) is much more common in salpingitis than in appendicitis. In early cases, a very high ESR may be a useful indicator, but its true diagnostic value has not been systematically investigated. Extreme pelvic tenderness and pain produced by motion of the cervix may occur in acute appendicitis, especially with rupture. Therefore, the finding should *not* exclude the diagnosis of appendicitis, as is so often thought. Although some feel that laparoscopy is of great value in making this distinction, hard data to support its safety and cost effectiveness, as well as its accuracy, are currently wanting.

Pyosalpinx. A pyosalpinx may rupture and cause a condition simulating pelvic appendicitis. The typical appendicular symptom sequence is usually wanting. If examination is made soon after the onset of symptoms, a pelvic swelling will be felt. This is usually bilateral. A history of chronic pelvic pain and leukorrhea may be ascertained. In patients coming under observation after some days, diagnosis may be impossible before operation.

Ovarian endometrioma. Rupture of an ovarian endometrioma may closely simulate a pelvic appendicitis. Sudden hypogastric pain, vomiting, slight fever, and tenderness on pelvic examination may be present in both cases, but with an endometrioma a unilateral or bilateral swelling ought to be detected on bimanual examination. A previous history suggestive of endometriosis is very helpful.

ULTRASOUND IN THE FEMALE

While a clear, positive diagnosis of a thick-walled appendix, and hence acute appendicitis, may not consistently be possible by ultrasound, this procedure is extremely useful in demonstrating an ovarian or tubal mass and the presence of fluid in the cul de sac. Remember, however, that after rupture and extrusion of the contents into the peritoneal cavity, a follicular or luteal cyst may no longer be detectable by ultrasound.

Late cases of appendicitis

In late cases of appendicitis that have led to a very diffuse or general peritonitis, or in those cases of a very fulminating type that are associated with a rapid form of spreading peritonitis, it is often impossible to make a certain diagnosis. Distinction has to be made from:

Primary pneumococcal peritonitis.
Secondary general peritonitis due to other causes (rupture of gastric, duodenal, typhoid, stercoral, or carcinomatous ulcer or of a pyosalpinx).
Thrombosis of mesenteric vessels.
Acute intestinal obstruction.
Acute pancreatitis.
Pylephlebitis.

In finding out the exact cause, the greatest importance attaches to the history. The subject is considered more fully in the next chapter and in that on peritonitis (Chapter 20).

8. Perforation of a gastric or duodenal ulcer

Perforation of a gastric or duodenal ulcer into the general peritoneal cavity is a catastrophe that often occurs with dramatic suddenness and, unless correctly treated, progresses until the death of the patient. It is one of the most easily diagnosed acute abdominal conditions, provided that the symptoms are known and appreciated, and it is the most important one to diagnose early and treat promptly, usually by surgical intervention. We say this advisedly, although we are aware that some patients have been and may be treated successfully without operation. Delay in the diagnosis of appendicitis is regrettable but does not always cost the patient's life. Misdiagnosis of an inflamed gallbladder or a pyosalpinx is a perilous occurrence, but the error may frequently be corrected by a later operation; but in the case of a perforated ulcer a delayed diagnosis or a misdiagnosis that leads to temporizing and delay is usually equivalent to a death sentence with a very slight chance of reprieve. If operation is undertaken within the first six hours, recovery is the rule; if the opening of the abdomen is delayed for twelve hours, recovery is more doubtful; if twenty-four or more hours elapse prior to suture of the ulcer the outlook is poor. True, some patients recover though operated on at a later stage than this, but they are always regarded as exceptional and worthy of comment. The very possibility of any condition's being due to a perforated ulcer is a positive indication for an *immediate* solution of the problem. If the problem cannot be solved with certainty, action should be taken for removal of the patient to the nearest surgical center. To leave the diagnosis in doubt overnight will most likely cost the patient's life if a perforation is present.

The signs and symptoms produced by the perforation vary according to the time that has elapsed since the rupture occurred. There are three stages in the pathological process that can usually be recognized easily, although there is no hard-and-fast limit between the stages.

The symptoms of each stage can be enumerated:

Early (within the first two hours)

Great and generalized abdominal pain.
Anxious countenance.
Livid or ashen appearance.
Cold extremities.
Cold, sweating face.
Subnormal temperature (95°F or 96°F).
Pulse small and weak.
Shallow respiration.
Retching or vomiting (slight).
Pain on the top of one or both shoulders.

Intermediate (two to twelve hours)

Vomiting ceases.
Abdominal pain less.
Appearance better, face regains normal color.
Temperature normal or slightly elevated.
Pulse normal.
Respiration still shallow and costal in type.
Alae nasi working slightly.

Abdominal wall very rigid, tender, and often retracted or flat.
Tender pelvic peritoneum.
Diminution of liver dullness.
Movable dullness in flanks (sometimes).
Great pain on movement of the body.

Late (after twelve hours)

Vomiting more frequent but still not profuse.
Facies of late peritonitis.
Abdomen tender and *distended.*
Pulse rapid and small; hypovolemic shock may be present.
Temperature usually elevated.
Abdominal wall usually not quite so rigid.
Respiration labored and rapid.

EARLY STAGE (FIRST FEW HOURS)

The initial symptoms are those due to the pain and the reaction
(probably due to the release of catecholamines) consequent on the
flooding of the peritoneal cavity with the gastric contents. The
sudden great stimulation of the innumerable nerve terminations by
the irritating fluid escaping from the ruptured viscus may be so
severe that the patient may feel faint or fall down in a syncopal
attack. The pulse temporarily is small and feeble, the face is livid,
the extremities are cold, and the thermometer may only register
about 95°F. The face shows pain and anxiety, and the patient may
cry out in his agony.

The pain is sudden in onset. The patient may be feeling well one
moment; the next, he is writhing in agony and crying out for
someone to relieve him. The site of the initial pain is generally
epigastric, but quickly it extends downward, and in a short time
it is felt all over the abdomen. The pain may even be greater in
the hypogastrium, since the escaped fluid collects in the pelvis.
This stage may last for only a few minutes or persist for an hour
or two. Its length depends to a certain extent upon the size of the

perforation and the degree to which the general peritoneal cavity is flooded. In cases where the perforation is very small and soon sealed up by fibrinous exudate, the symptoms of onset are correspondingly less severe. *In some instances this kind of response may be absent, especially if the perforation is small and the leakage minimal.*

INTERMEDIATE STAGE (TWO TO TWELVE HOURS)

The intensity of the initial pain subsides, and the patient then looks better and feels more comfortable. The circulatory system recovers to such an extent that the limbs may become warmer, the face normal in color, and the pulse normal in frequency and strength, while the thermometer may show no indication of either subnormality or fever. The improvement in symptoms does not imply any stoppage of the pathological process, though the casual observer might easily think that real improvement is taking place. Upon the proper appreciation by the practitioner of this dangerous latent period depends the patient's chance of recovery from the disease. It is in this stage that the inexperienced house surgeon thinks he has made a mistake in summoning the surgeon so urgently and almost apologizes for having brought him up needlessly. In this stage I have known a capable observer deluded into postponing the summons to the surgeon since the patient was sleeping peacefully. But it is at this period that the most favorable opportunity for operation should not be allowed to pass.

No certain guide is to be obtained from the pulse and temperature, for they are frequently normal, nor is the patient's own opinion of his condition always to be trusted, for he often expresses himself as feeling much better, and he may even begin to think lightly of his condition. But his attitude and his acts will always belie his words. Relief will be sought by the drawing up of the legs, and if he is asked to turn over in bed the attempt is made cautiously and with evident dread of increasing the pain. If no morphine has been administered there will still be a complaint of generalized abdominal pain, though the intensity will not be so great as at first. There are in addition five observations, some or all of which give valuable indications of the serious intra-abdominal mischief. The abdominal wall is rigid and tender, respiration is shallow and of the costal type, the pelvic peritoneum is tender, and there may be free fluid and free gas in the peritoneal cavity.

The rigidity of the abdominal wall is an almost constant feature. The muscles are flat and boardlike, and even firm pressure cannot make them give way. It takes a fairly deep anesthesia to cause them to relax. Presure on any part of the abdominal wall causes pain and may evoke retching. Tenderness is often greater in the right iliac fossa in the case of a ruptured duodenal or pyloric ulcer. The rigid muscles do not move on respiration, and the movement of the diaphragm is also considerably inhibited, so that *breathing is shallow and of the costal type.*

The *tenderness of the pelvic peritoneum* is a most important sign. This can be determined by a rectal or, in the female, by a vaginal examination. Within a very short time of a perforation the pelvis fills with escaped contents and inflammatory exudate, and though no lump can be felt, pressure against the pelvic peritoneal pouch through the rectal wall by the inserted finger produces pain that makes the patient wince (Fig. 19). Remember, however, that the tenderness of the pelvic peritoneum is not always present if the patient is examined within an hour or two of perforation and if the opening is small or quickly sealed up.

Movable dullness in the flanks due to free fluid in the peritoneal cavity should usually be determinable.

The diminution or absence of liver dullness is the sign produced by free gas in the peritoneal cavity. It is often easily demonstrated but is frequently ambiguous. Percussion over the front of the liver may produce a resonant note even when no free gas is present in the peritoneal cavity, for it may result from distended intestine that is sometimes pushed up in cases of intestinal obstruction or peritonitis from any cause. If there is no abdominal distention, however, diminution of the liver dullness anteriorly is significant. It is always of significance to obtain resonance on percussion over the liver in the midaxillary line. *If in any acute abdominal case distinct*

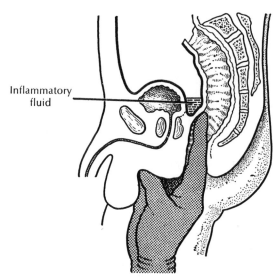

Inflammatory
fluid

Fig. 19. How the pelvic peritoneum may be palpated and indicating where inflammatory fluid usually collects.

resonance is obtained over the liver in the midaxillary line about two or more inches above the costal border, one is certainly dealing with a perforation of a gastric or duodenal ulcer (Fig. 20). It is only in a minority of cases that the sign is positive, however.

An additional symptom that may be helpful is pain on top of the shoulder, in the supraspinous fossa, over the acromion, or over the clavicle, that is, in the region of distribution of the cutaneous branches of the fourth cervical nerve. This symptom, if present, has to be considered carefully with the other indications, for diaphragmatic pleurisy causes similar pain; but if the pain is felt on both shoulders *from the onset of the attack*, it is suggestive of a perforation of the anterior wall of the stomach, causing irritation of the median portion of the diaphragm. In the case of a perforation of a pyloric or duodenal ulcer, the shoulder pain is usually felt in the right supraspinous fossa.

It is possible to get great help from a simple radiograph of the diaphragmatic region. By this means, small quantities of free gas

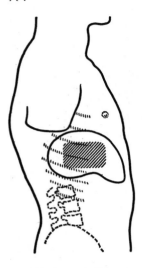

Fig. 20. The area of hepatic resonance that is diagnostic of perforation of a gastric, duodenal, or intestinal ulcer.

between the liver and diaphragm may be observed (Plate 2), but in 15–20 percent of the cases of perforated ulcer, no free gas can be demonstrated. In some instances in which the diagnosis is in doubt even after a careful history, examination, and plain films of the abdomen, the oral administration of Gastrografin may demonstrate a perforation or exclude its presence. Such studies are not often necessary but may be helpful in late cases, especially if the question of sealing of the perforation has arisen.

LATE STAGE (AFTER TWELVE HOURS)

Locally the extensive peritonitis is clearly shown by increasing *distention of the abdomen.* This distention is not a sign of a perforated ulcer; it is an indication that peritonitis is advanced and that the condition has been allowed to proceed too far. Yet I have known delay in sending a patient to the hospital because the doctor who suspected perforation thought that the diagnosis could hardly be sustained in the absence of distention.

The other effects of extensive peritonitis are increasing and persistent vomiting, a gradual increase in rate and a decrease in the force and volume of the pulse, and a consequent decrease in the temperature of the extremities and the body generally. The abdomen remains tender, but in late peritonitis the rigidity frequently lessens. Finally, as a result of the vomiting and depressed circulation, the face becomes pinched and anxious, the cheeks hollow and the eyes dim and beringed with dark circles, the so-called *facies Hippocratica*, which is not so much a sign of peritonitis as the mask of death following peritonitis. The more common causes of severe peritonitis and acute collapse are shown in Figure 21.

DIAGNOSIS

During the initial stage it is nearly always possible to say that there is a condition needing surgical intervention, though the exact nature of the catastrophe may be slightly doubtful. Great help is sometimes obtained from a previous history of chronic indigestion or of duodenal pain coming on about two hours after taking food. Quite a number of patients, however, give only a recent history of pain after taking food. This is more common in young people in whom acute pyloric ulcers appear to be not uncommon, but I have had to deal with a perforated duodenal ulcer in a man of seventy-six who had never had any symptoms of indigestion.

If, in a patient who has been subject to chronic indigestion, sudden collapse and very severe abdominal pain suddenly supervene, and if at the same time the abdominal wall becomes generally rigid, one is justified in suspecting perforation of an ulcer. If, in addition, the pelvic peritoneum is tender and there is resonance over the lateral aspect of the liver, the diagnosis is certain.

In the second stage the general symptoms temporarily improve, but *all the local signs remain and become still more definite*, so that the careful observer should not be misled. In the third stage, there is no difficulty in diagnosing that some serious catastrophe within the abdomen has occurred.

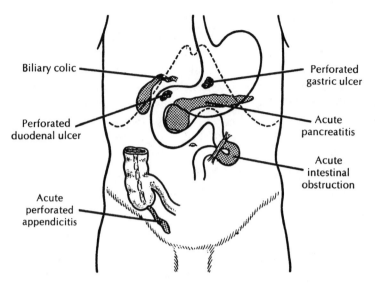

Biliary colic

Perforated
duodenal ulcer

Acute
perforated
appendicitis

Perforated
gastric ulcer

Acute
pancreatitis

Acute
intestinal
obstruction

Fig. 21. The more common abdominal causes of acute collapse. (In the female
ruptured ectopic gestation should be added.)

The finding of free peritoneal air or the leakage of water-soluble
contrast into the peritoneal cavity are useful findings in doubtful
cases.

DIFFERENTIAL DIAGNOSIS

There are three conditions, sometimes giving rise to symptoms
similar to those of perforated ulcer, that either do not call for
operation or in which operative interference is positively contraindicated. They are:

1. Severe colic (either biliary or renal).
2. Some cases of pleuropneumonia or pulmonary infarction.
3. Gastric crises of tabes dorsalis.

Colic. Biliary and renal colic may cause severe collapse and terrible
abdominal pain. The extent of the collapse is not a differential

point, since in biliary colic the patient may sometimes appear *in extremis*, but the diagnosis is usually clear on consideration of the patient's history and on careful observation of the condition of the abdominal wall, the liver dullness, and the pelvic peritoneum. A clear account of prior attacks of pain and jaundice, or hematuria and the passing of gravel or a small stone, would indicate the probability of a stone trying to pass down the biliary ducts or ureter, respectively. The radiation of the pain of biliary colic to the subscapular region, and that of renal colic to the testicle, are sufficiently diagnostic. In ureteral stone colic the abdominal wall is not usually rigid, and the sufferer may throw himself about or writhe in agony while attempting to gain a more easy position. After perforation of an ulcer the general abdominal rigidity and increase in pain on movement forbid and prevent movement, though occasional exceptions occur. Finally, the pelvic peritoneum is not tender, nor is there any diminution of liver dullness in biliary or renal colic. If jaundice or hematuria is observed, the diagnosis will not be in doubt. *The pain of renal colic is nearly always limited to one side.*

Right-sided or bilateral pleuropneumonia or pulmonary infarction. An acute case of bilateral or right-sided pleuropneumonia or pulmonary infarction will sometimes cause considerable abdominal rigidity and great epigastric pain, but in such cases there are usually sufficient signs in the lung to point to the true cause of the condition. The alae nasi will be working, and the respiration rate will be greater than one would expect with an early peritonitis without distention. With pleuropneumonia there is usually fever and a raised pulse rate. Once more, rectal examination and percussion over the lateral aspect of the liver are of importance in diagnosis.

Tabes dorsalis. The gastric crises of tabes dorsalis (*now rarely seen*) may give rise to difficulty in diagnosis, for the intensity of the abdominal pain and the severity of the vomiting may cause extreme collapse. It should be a rule always to test the knee jerks and the pupillary reactions in every acute abdominal case, for in tabes one or the other of these is nearly always abnormal. *Persisting rigidity of the abdominal wall is never due to tabes*, and tragic misdiagnosis may occur if this point is not remembered.

There are six other conditions that are sometimes difficult to distinguish from a perforated gastric or duodenal ulcer and that are more commonly confused with perforated ulcer than the above three conditions. They are:

1. Acute pancreatitis.
2. Acute perforative appendicitis.
3. Ruptured ectopic gestation (in women).
4. Acute intestinal obstruction.
5. General or diffuse peritonitis from other causes.
6. Postemetic rupture of the esophagus.

To these should be added the rarer conditions of mesenteric thrombosis and dissecting aneurysm. When the diagnosis of acute pancreatitis is quite clear, most surgeons now reserve operation until the acute symptoms have subsided and special indications appear.

Acute pancreatitis. This simulates visceral perforation very closely, and before the abdomen is opened it is quite often mistaken either for that condition or for intestinal obstruction. In pancreatitis the pain may be even more agonizing, but the abdominal rigidity is not so generalized or so constant. Cyanosis and slight jaundice are more often seen in pancreatitis, which often occurs in obese patients. The diagnosis is further considered in Chapter 9.

Acute appendicitis. This should easily be distinguished by consideration of the history, the order of the symptoms, and the local signs. Inflammation of the appendix infrequently causes such acutely severe symptoms as those ushering in a gastric perforation, but in the second stage, a perforated ulcer may be, and often is, misdiagnosed as appendicitis. Especially is this the case with a leaking duodenal ulcer, for the escaped contents may trickle down chiefly on the right side of the abdomen and cause pain, particularly in the right iliac fossa. This simulates appendicitis closely, for the sequence—epigastric pain, nausea and vomiting, right iliac pain and fever—may be produced just as in inflammation of the appendix. The intensity of the initial collapse may serve to distinguish, and the persistence of tenderness over the duodenal area should help to determine, the condition. In appendicitis the abdominal rigidity is seldom as extensive as in perforated ulcer, and the liver dullness is normal, though in both cases there may be rectal tenderness. In many cases of perforated duodenal ulcer, the patient may complain of pain on the top of the right shoulder or over the right supraspinous fossa; very rarely is shoulder pain felt in appendicitis, and when it occurs (due to irritative fluid reaching the diaphragm), the pain would be more likely felt over the acromion or the clavicular region. In both cases operation is indicated.

Intestinal obstruction. This should not give rise to difficulty except in those patients who come late for diagnosis. Acute strangulation of a coil of small bowel is the most likely type to cause difficulty. In both conditions the onset may be with acute symptoms of collapse, pain, and vomiting, and in both there may be evidence of free fluid in the abdomen, but in obstruction the abdominal wall is usually flaccid and quite unlike the rigid, boardlike condition in perforated ulcer. In acute obstruction vomiting is almost from the first a distinctive feature, and the character of the vomit gradually changes until it is feculent. If, however, a coil of small intestine is acutely strangulated and lies in contact with the anterior abdominal wall, there may be tenderness on pressure and sometimes rigidity of the overlying muscle.

In the late stages of both conditions it may be difficult to distinguish between them, for peritonitis is often a complication of late intestinal obstruction, and the boardlike rigidity accompanying a perforated ulcer tends to diminish somewhat as the distention increases. In such cases the history, and possibly the character, of the vomit may serve to differentiate these conditions.

Ruptured ectopic gestation. Rupture of an ectopic gestation, leading to severe intraperitoneal hemorrhage, may cause syncope and collapse, vomiting, and severe abdominal pain. A history of menstrual irregularity may be obtained, but one must not rely on that for diagnosis. The main points in diagnosis are the blanching of the lips, tongue, nails, and sclerotics and the absence of true abdominal rigidity, though the abdomen is generally tender and tumid, especially in the lower part. In both cases there will be some tenderness on digital rectal or vaginal examination. No definite pelvic swelling can be made out in most cases of recent rupture of an ectopic gestation. Free fluid in the abdomen may be detectable in both conditions, and resonance over the *front* of the liver may be obtainable sometimes with a ruptured ectopic gestation (due to intestine pushed up by clots of blood), but resonance over the lateral aspect of the liver is obtained only by a perforated ulcer.

Pain over the clavicles or in the supraspinous fossa is sometimes a complaint in cases of ruptured extrauterine gestation, as with perforated ulcer. This is due to diaphragmatic irritation by the clotted blood in the upper abdomen.

Dissecting aneurysm. A dissecting aneurysm of the aorta may cause acute abdominal pain, collapse and rigidity of the muscles of the abdominal wall, and has on occasion given rise to a diagnosis of a perforated peptic ulcer. However, the thoracic origin of the pain, the absence of free gas in the peritoneal cavity, and the reduction or absence of the pulse in one of the lower limbs should serve to differentiate the condition.

Peritonitis from other causes. Other forms of peritonitis can only be distinguished from that due to a perforated ulcer by considering the history of onset and by determining the presence or absence of gas on the lateral aspect of the liver. It may be impossible to differentiate from the results of perforation of some other part of the gut. Peritonitis due to rupture of the gallbladder may be accompanied by an icteric tinge in the conjunctivae.

Postemetic rupture of the esophagus. The history is the most important differentiating feature because the pain of this condition almost invariably follows forceful retching and vomiting. Occasionally, after the initial retching, hematemesis may be evident just prior to the development of the pain. The pain is high in the epigastrium and often radiates through to the lower thoracic spine. It is usually aggravated by lying flat and is relieved somewhat by sitting. Dyspnea is relatively common, and crepitus may be detected in the cervical and supraclavicular regions. The abdomen may appear to be somewhat guarded, but true rigidity and rebound tenderness are absent. Evidence of a pneumothorax or pleural effusion may be found by physical examination and on the X-ray. The diagnosis of postemetic rupture is relatively simple if it is considered, and it can be confirmed by films taken while the patient swallows a water-soluble dye.

Rupture of an ulcer with formation of localized subphrenic abscess

When for one reason or another—previous adhesions, slow leakage allowing time for deposition of fibrin—the escaping gastric contents do not flood the peritoneal cavity, the symptoms are correspondingly modified. The pain may be very great but the initial collapse is not so prostrating, and the abdominal signs will soon be localized to the upper segment of the abdomen and lead to the development of a subphrenic abscess containing gas. If such an abscess develops anteriorly, the local signs of intraperitoneal suppuration are very evident, but when the mischief is high up under the diaphragm, the signs and symptoms take longer to develop. Irregular temperature, rigors, leukocytosis, and dullness at the base of the lung consequent on pleural effusion or basal congestion will lead the observer to diagnose a collection of pus under the diaphragm. A full description of subphrenic abscess does not come within the scope of this book. It must be remembered that occasionally a duodenal or pyloric ulcer may perforate, but the perforation may soon be sealed by a fibrin deposit. Such cases give rise to pain, rigidity, and tenderness in the right hypochondrium closely simulating the symptoms of *acute cholecystitis*, and the condition may clear up without the formation of an abscess. On the other hand, such a sealed perforation may become unsealed and cause a recrudescence of the acute symptoms.

Perforation into the lesser sac. Frank perforation of an ulcer into the lesser sac of the peritoneum is not common. Usually adhesion takes place to the posterior wall of the sac, and the ulcer simply erodes into the pancreas. When, however, frank perforation does take place, there is initial acute pain in the upper abdomen, but rigidity of the upper abdomen does not result unless the fluid escapes from the foramen of Winslow into the general peritoneal cavity. An abscess in the lesser sac is likely to follow, giving rise to fever and the usual signs of suppuration and forming one type of subphrenic abscess.

9. Acute pancreatitis

Acute pancreatitis seems to be increasing in incidence, possibly because of the burgeoning use of alcohol. In hospitals with large indigent populations, roughly two-thirds of the cases are associated with alcoholism, whereas in private institutions, two-thirds of the cases are associated with gallstones. The passage of small stones into the duodenum is probably an important causal factor since stones have been demonstrated in the stools of patients with pancreatitis even though major obstruction of the ampulla of Vater by a stone is rarely encountered. After these two major inciting factors (alcohol and gallstones), about 10 percent of cases can variously be attributed to trauma, hyperlipidemia, hyperparathyroidism, and drugs such as thiazide diuretics. In most reports, roughly 10 percent of cases cannot be classified according to cause.

It has been stated that the common failure to diagnose acute pancreatitis correctly is due to a failure to consider its possibility in the individual case, but even when the condition is thoroughly considered and discussed, a mistaken diagnosis frequently results. A special consideration of the symptoms is therefore all the more necessary.

To understand and remember the symptoms, one should recollect the antomy of the pancreas and the pathology of the disease.

123

The gland lies in the retroperitoneal tissues in close relationship with the celiac plexus and ganglia. The head is surrounded by and slightly overlaps the duodenum; the body lies in front of the first lumbar vertebra, while the tail reaches the left loin and lies against the hilum of the spleen. There are still many points in the pathology of pancreatitis that are not settled, but there is a preponderance of evidence to show that the acute forms of inflammation may lead to severe, widespread hemorrhage into the gland and surrounding tissues, with subsequent disorganization of its substance and liberation and activation of its ferments. Pancreatic enzymes cause extensive destruction of *retroperitoneal tissues*. If a major pancreatic duct has been disrupted, a collection of pancreatic juice forms, bounded by whatever tissues happen to be in the area, and is called a *pseudocyst*. Less commonly, there is a free communication of the disrupted duct into the peritoneal cavity, giving rise to *pancreatic ascites*. Bacteria probably play no role in the early onset of the pancreatic inflammation but tend to cause secondary abscesses in the retroperitoneum about two weeks after the beginning of the attack. If the patient lives long enough, a part or the whole of the pancreas may become gangrenous, but this is very rare.

Acute pancreatitis seldom occurs before the age of twenty and is more common in stout people.

Symptoms

We describe a *severe* case. In milder cases the symptoms are correspondingly less acute.

The symptoms of acute pancreatitis are rather variable—a fact that explains the conflicting accounts of the disease published by individual observers. The one or two absolutely pathognomonic signs are rarely present because they are found late in the course of the disease, and the more constant features must be carefully considered together before a diagnosis can be determined. It is better to group the manifestations according to their cause.

SYMPTOMS AND SIGNS PROBABLY DUE TO INFLAMMATION
OF THE GLAND

Pain. Though there may have been slight attacks of pain prior to the
main attack, the acute onset is usually dramatically sudden, and
fainting may occur, but occasionally the onset is insidious. The pain
is usually excruciating and the patient will cry out in agony. The
pain is felt in the epigastric zone and *in one or both loins*, though it
may be present only in the left hypochondrium. The position of the
gland accounts for the loin pain, and the neighborhood of the celiac
plexus explains its severity. Sometimes pain is felt in the left
scapular region and occasionally in the left supraspinous fossa
(phrenic pain). Later the intensity of the pain diminishes, but it
may be felt over the whole abdomen or perhaps more *in the right
iliac fossa.*

Shock. Shock usually accompanies the pain. The cold extremities,
sweating skin, and weak pulse are sufficient witness to the severity
of the shock. Enormous volumes of intravenous fluid may be
required to maintain an adequate circulation because the extreme
insult to the tissues exposed to the enzymes causes a marked
outpouring of plasma volume and extracellular fluid. The temperature may be as low as 95°F initially but soon rises, often remarkably
rapidly and depending upon the severity of the attack. The pulse is
usually rapid and weak.

Vomiting. Reflex vomiting or retching nearly always occurs. Sometimes the retching is incessant, but as a rule very little material is
brought up. The vomiting is more persistent than with a perforated
ulcer. Occasionally no nausea is felt. In true reflex vomiting the
vomit is never feculent.

Fever. It is a little known fact that fever occurs almost invariably in
bona fide cases. It may reach 102°F or 103°F, but a rise even to
104°F *early* does *not* signify the presence of a pancreatic abscess.
On numerous occasions we have seen experienced surgeons oper-

ate for pancreatic abscess early in an attack because of fever and leukocytosis, only to find none.

Tenderness. Local epigastric tenderness is a constant finding.

Rigidity. Epigastric rigidity is by no means constant. It is true that soon after the onset there may be boardlike rigidity of the epigastric muscles, but when the patient is examined there is *often a lax abdominal wall.* This point should be emphasized, since extreme muscular rigidity was at one time thought to be characteristic.

SYMPTOMS AND SIGNS DUE TO ENLARGEMENT OF THE PANCREAS

Epigastric tumor. Sometimes the pancreas may be palpable as a transversely placed tumor in the epigastrium. The fact that the patient is sometimes very stout and the occasional presence of rigidity often make the detection of the tumor difficult.

Jaundice. Slight jaundice is found in about half the cases. Since frequently, if not usually, there are no obstructing gallstones, the most reasonable explanation for the jaundice is that the common duct is compressed by the swollen head of the pancreas.

Obstructive vomiting. True obstructive vomiting of feculent or bilious material is very rare, but I have personal knowledge of one such case. At the operation the swollen pancreatic head was definitely obstructing the duodenum. Obstructive vomiting must be distinguished from the more common reflex vomiting mentioned above.

SIGNS DUE TO EXTRAVASATION OF BLOOD

Ecchymosis. In hemorrhagic pancreatitis some of the blood may find its way along the tissue planes and become evident as a green or greenish-yellow area in one or both loins. This was first noted by Grey Turner, who also, in one patient, saw a similar discoloration around the navel (sometimes called Cullen's sign). This symptom, when found, is pathognomonic of retroperitoneal hemorrhage, but it never occurs until two or three days after the onset of the disease. It usually signifies the presence of a rather severe form of the disease.

OTHER SYMPTOMS AND SIGNS

Cyanosis. This symptom has been noted in a considerable number of cases. It is best observed in the face and extremities but has sometimes been present in the skin of the abdomen.

Tachypnea and dyspnea. These are often noticeable. It is reasonable to suppose that a partial inhibition of the diaphragmatic movements, owing to the contiguous inflammation, may account, at any rate in part, for the cyanosis and dyspnea. A left pleural effusion is very common and likewise contributes to these symptoms, as does the sometimes extreme abdominal distention. Some have suggested that circulating lipases may destroy pulmonary surfactant and hence produce atelectasis and pulmonary failure.

It should be remembered that in the later stages of acute pancreatitis a more general abdominal condition results; blood-stained fluid collects in the peritoneal cavity, distention supervenes, and there may be persistent high fever. It is very difficult to diagnose such cases without a very accurate previous history of the case. Very rarely indeed subcutaneous areas of fat necrosis may develop in pancreatitis.

LABORATORY FINDINGS

Hyperglycemia and glycosuria. These are sometimes found, and in any case of acute abdominal pain should raise the question of pancreatic disease.

Increase in the serum and urinary amylase. The liberation of the pancreatic ferments leads to an increase in the amount of amylase in the blood and the urine.

The serum amylase is less than 200 Somogyi units per 100 ml of

serum in about one-third of cases; between 200 and 500 units in
one-third, and greater than 500 in the remaining third. If an
elevation is present, it is usually found within the first forty-eight
hours, returning to normal at that time even if the disease remains
active. However, inasmuch as the amylase content of the serum
may be significantly raised in patients suffering from cholecystitis,
high intestinal obstruction, acute renal insufficiency, perforated
ulcer, and other conditions, the serum amylase is not as reliable a
test for acute pancreatitis as it was at one time supposed to be.
While determination of the urinary amylase improved the diagnostic accuracy somewhat, it, too, fell short of the mark. Recently it has
been demonstrated in cases of pancreatitis that there is a discordant
elevation in clearance of amylase by the kidneys in comparison with
the creatinine clearance so that the ratio of amylase:creatinine
clearance exceeds 5:1. Great hopes were held for this test but, as is
so often the case, many exceptions have been found and high ratios
have already been reported in burn patients and in individuals with
diabetic ketoacidosis. It is probably of no greater value than a serum
or urinary amylase or lipase test.

Hyperlipidemia. It is thought that primary hyperlipidemia may
account for some cases, but so-called secondary hyperlipidemia is
much more common. *A lipemic serum in a seriously ill patient with
abdominal pain almost always means pancreatitis.*

Other laboratory findings. Albuminuria is common. If methem-
albumin is detected in the serum of a patient suffering from acute
pancreatitis, it is a case of hemorrhagic pancreatitis.

Diagnosis

Acute pancreatitis is most commonly mistaken for a *perforated
gastric* or *duodenal ulcer.* The less acute cases may be misdiagnosed as *appendicitis,* while those cases with distention may easily

be regarded as examples of *intestinal obstruction*. *Acute chole-cystitis* and *biliary colic* may also simulate the symptoms of pancre-atitis.

With a perforated ulcer, general abdominal rigidity is constant in the early stages after perforation, while in pancreatitis the abdomen may be softer, and any rigidity is usually limited to the epigastric zone. In his original paper, Fitz very accurately wrote that the symptoms of acute pancreatitis were those of an *epigastric* peri-tonitis. In a case of perforated ulcer the symptoms are usually more widespread. Pain on top of the shoulder is frequently felt when an ulcer perforates; with pancreatitis such pain is rare, and when present is felt on top of the *left* shoulder. Bilateral lumbar pain, cyanosis, fever to 102°F or 103°F, and slight jaundice would favor pancreatitis, while absence of liver dullness in the axillary line would definitely indicate perforated ulcer. Hyperglycemia, gly-cosuria, hyperlipidemia, or an increased serum or urinary amylase would point to pancreatitis.

Appendicitis is generally distinguishable if careful attention is paid to the history of onset and the order of symptoms. The vomiting and pain are both less severe in appendicitis, and there may be definite local symptoms in the right iliac fossa. With acute cholecystitis and biliary colic tenderness is felt more in the right hypochondrium, and there may be a definite history of previous attacks. The laboratory tests may help to determine the condition, but it must be remembered that cholecystitis and pancreatitis may coexist.

When distention has supervened, pancreatitis is difficult to distinguish from the late stage of peritonitis and intestinal obstruc-tion unless positive laboratory tests and a very clear history point to the correct diagnosis.

Perforation of the duodenal diverticulum, although not common, may give rise to a clinical picture indistinguishable from that of acute pancreatitis. The pain and tenderness are usually more localized to the right upper quadrant. Films may help to distinguish the condition by demonstrating gas bubbles in the right retro-peritoneal area.

Abdominal paracentesis may yield a prune-juice-colored fluid with a very high amylase content, but so may strangulation obstruction of the intestine.

With every care in investigation, acute pancreatitis is sometimes only diagnosed with certainty when the abdomen is opened and bloodstained fluid and areas of fat necrosis are seen.

PSEUDOCYST: A COMPLICATION OF ACUTE PANCREATITIS REQUIRING INTERVENTION

This complication of acute pancreatitis should be suspected if after having come through the early phase successfully, the patient does not progress rapidly to full recovery. There is usually persistent fever, low-grade abdominal pain, especially after eating, anorexia, and occasional vomiting. A mass is sometimes palpable in the epigastrium. The serum amylase level, rather than returning to normal, as in the usual case, remains persistently high. A CT scan will show a well-defined cystic collection whose resolution or progression is best determined by serial CT examinations. Whether a pseudocyst is infected or sterile can only be determined by aspiration of the fluid under CT or ultrasound guidance. Some pseudocysts will resolve spontaneously, while others require drainage.

"PANCREATIC ABSCESS"

This term is a misnomer because the condition is neither an abscess in the true sense of the word nor is it confined to the pancreas. Rather, there is extensive destruction of the peripancreatic and retroperitoneal fat accompanied by secondary infection, usually with gram-negative rods. These retroperitoneal tissues do not coalesce into an easily drained abscess cavity but require active debridement.

The clinical setting and findings are identical to those described above for pseudocyst, except that the CT scan shows multiple poorly perfused areas of pancreatic and peripancreatic tissues without the well-defined collection seen with a pseudocyst. The diagnosis should be considered in any patient with acute pancreatitis who has persistent abdominal pain and fever. Prompt diagnosis is extremely important because the condition is *uniformly fatal* without surgical intervention.

In cases of nonresolving acute pancreatitis, the diagnosis is established with certainty if gas can be detected in the retroperitoneum by either a plain film or CT scan, if there is persistent bacteremia without any other cause, or if CT-guided aspiration of the peripancreatic and pancreatic areas yields bacteria.

Some surgeons have advocated debridement when pancreatic and peripancreatic necrosis without superimposed infection is present. The diagnosis of necrosis alone (in absence of infection) has been based on poorly perfused areas shown on a contrast-enhanced CT scan. Whether these areas are in fact necrotic pancreas, and whether an operation should be undertaken for them, is extremely doubtful in my opinion.

10. Cholecystitis and other causes of acute pain in the right upper quadrant of the abdomen

Severe pain arising in, or chiefly localized to, the right hypochondriac region is usually due to one of the following conditions:

1. Cholecystitis, with or without rupture.
2. Biliary "colic" (a misnomer).
3. Inflamed or leaking duodenal ulcer.
4. Hepatitis.

But one always needs to exclude:

1. Appendicitis.
2. Renal pain or colic.
3. Pleurisy or pleuropneumonia.

The gallbladder and cystic duct may be regarded as a vermiform muscular tube that has a dilated extremity and opens through the common bile duct into the duodenum. In certain respects, therefore, it is analogous to the cecal appendix. Further, it is common for a stone to stop up the cystic duct just as a concretion may occlude the lumen of the appendix. The chief difference between the two structures lies in the fact that fecal material is present in the cecal appendix but not in the gallbladder, though *Bact. coli (Escherichia coli)* is frequently found in the latter. In addition, the muscular wall of the gallbladder and common bile duct is sparse, so that vigorous contractions do not occur as they do in the ureter or in the intestine.

Acute cholecystitis

PRODROMAL STAGES

The usual attack of acute cholecystitis begins with an episode of biliary colic (see Chapter 12). Since this pain is almost always without paroxysms or with only very minimal ones, one should not be misled by the term biliary colic or by the patient with *steady* pain. An attack of biliary colic may follow one of four courses:

1. Usually it subsides over four to six hours because the calculus that caused the attack has either dropped back into the gallbladder or has passed out of the common bile duct.
2. Acute cholecystitis may supervene, often after an interval of several hours of freedom from pain.
3. A chronic hydrops of the gallbladder may form after impaction of the stone in the cystic duct, provided that infection is not present and the mucosa of the gallbladder does not undergo necrosis.
4. Jaundice may occur transiently if the stone has passed into the duodenum or does not remain critically obstructive of the common bile duct. If it does not pass and partial obstruction of the common bile duct persists, then cholangitis may ensue.

It is to be emphasized that *an episode of biliary colic causes steady, nonparoxysmal pain* and that it may be the forerunner of any of the above-mentioned events. *Pain arising from a stone in the cystic duct or common bile duct is essentially the same because of identical segmental innervation of these structures.* Consequently, during the painful attack itself, the surgeon is rarely, if ever, able to predict which course of events outlined above will ensue. Occasionally acute cholecystitis occurs in the absence of gallstones or obstruction of the cystic duct, but this accounts for only about 5 percent of cases.

SYMPTOMS, SIGNS, AND DIAGNOSIS

The main symptoms and signs are:

1. Pain.
2. Vomiting or nausea or complete loss of appetite.
3. Jaundice (sometimes).
4. Fever.
5. Palpable or visibly distended gallbladder.
6. Tenderness and rigidity.

Pain. The pain of biliary colic is described in Chapter 12. One point bears emphasis here: its constant, unrelenting nature and its very frequent *midline* epigastric position often mislead the physician. The acute pain usually disappears within four to six hours. After a pain-free interval of several hours, it returns, but now in the right hypochondrium and with a different character, that of a constant ache clearly aggravated by a motion, cough, or sneeze. The secondary pain is clearly that of peritoneal irritation and is analogous to the shift from visceral to somatic pain in acute appendicitis.

Vomiting. Anorexia, nausea, and vomiting are not only common in biliary colic but usually persist when acute cholecystitis is present. Vomiting is usually rather frequent and is bilious, not feculent, in character. Vomiting is occasionally completely absent.

Jaundice. A slight tinge of icterus is quite common in acute cholecystitis, even in the absence of stones in the common bile duct. Occasionally its presence is helpful in differentiating acute cholecystitis from a perforated ulcer. The more severe the attack of acute cholecystitis, the more likely the presence of icterus. Serum bilirubin levels of 3–4 mg percent are not at all uncommon in uncomplicated acute cholecystitis, and I have encountered one case which reached 11 mg percent.

Fever. Fever of about 101°F is common in acute cholecystitis. Higher fever or rigors, while occasionally observed in acute cholecystitis, are more likely to occur in cholangitis.

Palpable or visibly distended gallbladder. A palpable gallbladder has been reported in close to half of all cases of acute cholecystitis. The *presence of a palpable gallbladder in a patient with a compatible history establishes the diagnosis of acute cholecystitis.* A rough examiner who palpates too deeply will miss this finding because the patient will guard against the probe. We have often *seen* a distended gallbladder descend on deep inspiration while standing at the foot of the bed with the patient in a good light when the same turgid organ has been missed on palpation by an untrained, rough hand. Gentle palpation will often reveal a very hard mass slightly larger than a golf ball that cannot be mistaken for anything else once a few have been felt. Millions of dollars in diagnostic tests could be saved this way. With the rare exception of the tiny, shriveled intrahepatic gallbladder, years of operative experience indicate that the vast majority of gallbladders in acute cholecystitis are markedly distended and that once anesthesia is induced, most are palpable. This suggests that if we train ourselves to be gentle, most gallbladders in acute cholecystitis will probably be palpable.

DIFFERENTIAL DIAGNOSIS

Cholecystitis must be distinguished from appendicitis, an inflamed or leaking duodenal ulcer, biliary colic, rupture of the gallbladder or bile duct, and hepatitis, but pleurisy must always be excluded (Fig. 22).

The symptoms—pain, vomiting, constipation, fever—are very similar to those of *appendicitis*, but the site of localized pain is in the

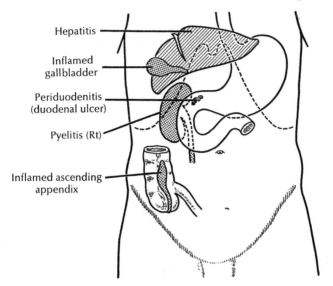

Hepatitis

Inflamed
gallbladder

Periduodenitis
(duodenal ulcer)

Pyelitis (Rt)

Inflamed ascending
appendix

Fig. 22. The differential diagnosis of cholecystitis.

one case in the right hypochondrium and in the other in the right
lumbar or iliac region. If any swelling is palpable, its continuity
with or distinction from the liver is of prime importance in diagnosis. It must be allowed that there are some cases, especially in
obese patients with a rather low-lying, inflamed gallbladder accompanied by local peritonitis and rigidity, in which a definite differentiation from appendicitis with local abscess is almost impossible
before operation. A history of jaundice or biliary colic may be of
assistance, or an account given of previous attacks suggestive of
appendicitis might point to that disease.

In some cases of acute cholecystitis the initial pain may be felt in
the epigastrium and may precede by several hours the pain and
tenderness in the right upper quadrant of the abdomen. This is
comparable to the epigastric or umbilical pain that may precede the
right iliac pain and tenderness in cases of acute appendicitis. All
patients suffering from acute epigastric pain unaccompanied by any

other localizing symptom should be examined again within a few hours.

In difficult cases in which the gallbladder cannot be palpated, intravenous cholangiography or a radionuclide scan (HIDA) is useful. Visualization of the gallbladder excludes the diagnosis of acute cholecystitis. Recent experience indicates that ultrasound examination of the gallbladder may demonstrate calculi that cannot be shown by other methods (Plate 4), but the degree of distention of the gallbladder is difficult to ascertain by ultrasound.

CASE REPORT OF THE AUTHOR'S ATTACK

(Z.C.)

The only previous abdominal crisis I had suffered was an attack of acute gangrenous appendicitis in 1907; an abscess formed and was drained, and later the stump of the appendix was removed at another operation.

For six months before my illness I occasionally had attacks of profuse sweating and increased pulse rate that woke me from sleep, but usually passed off within half an hour. They were not accompanied by any pain. A physician found my heart normal.

On 3 April 1969, I arose about 7 a.m., enjoyed a simple breakfast, and was up and about all the morning. Then, almost exactly at noon, I suddenly felt a dull severe, deep pain in the middle of the epigastrium. I lost all appetite for food and retired to bed. I wondered what might be the cause of the pain, and thought first of coronary thrombosis, but the pulse was normal in every respect, and there were no symptoms to support this diagnosis. My attention was then directed to the abdomen. I palpated the left side and the right iliac region without finding anything abnormal. In the right hypochondrium, however, a surprise awaited me, for in the normal position of the gallbladder was a rounded, tense, and firm swelling, about the size of a small golfball. It was not painful or tender. This absence of tenderness probably put all thoughts of acute cholecystitis out of my mind. I decided to phone an experienced medical practitioner, and he arranged to visit me that afternoon. As I rested in bed the epigastric pain became easier and I dozed a little on and off, and did not even once feel if the abnormal swelling had changed. The physician arrived later in the afternoon, examined the abdomen very carefully, but found no abnormal swelling in the gallbladder area, nor anywhere else in the abdomen. This greatly surprised me, but I was able to confirm that the swelling I had felt had now completely disappeared. Both pain and swelling having now gone it was agreed that there was no need for any special line of treatment. I remained comfortable and free from pain for four to five hours. But about 9 p.m. I began to have pain in the right hypochondrium and tenderness in the same area. The pain increased in severity, so I phoned the physician, told him of the change, and asked him to pay me another visit on the morrow. The second visit was paid on 4 April. When the physician saw the position of the pain and tenderness, and found I had a slight rise of temperature, he promptly called into consultation a distinguished surgeon, who confirmed the diagnosis of acute cholecystitis and advised my immediate removal to hospital. When I arrived there the diagnosis was again confirmed by a surgeon of vast experience, and he advised immediate operation. This took place that same evening, and a diseased gallbladder, already necrotic in parts, and containing 15 pigment stones, was successfully removed. Convalescence was uneventful, and I was discharged on 18 April.

COMMENT

The course of events was probably as follows:

Bacterial action first attacked the mucosa, which responded by a copious secretion of mucus that distended the viscus and caused the epigastric pain. The distension may have been due to the inability of the weak muscle fibres in the wall of the viscus to push the sticky mucus up through the narrow opening of the cystic duct, or maybe a small stone had temporarily stopped up the opening. In either case lying horizontally in bed permitted the mucus to escape from the gallbladder, relieved the distension, and eased the epigastric pain. During the four to five hours of freedom from pain bacterial action slowly penetrated the whole thickness of the wall of the viscus, and thereupon irritation of the parietal peritoneum led to pain and tenderness in the right hypochondrium. The next stage would have been perforation of the wall of the gallbladder and diffuse peritonitis.

This experience taught me with regard to acute cholecystitis (1) that the first symptom may be epigastric pain, (2) that the first sign may be a distended but not tender gallbladder, (3) that for one reason or another the gallbladder may then be able to expel its contents and both swelling and pain may disappear, (4) that a period of four or five hours may then ensue in which the patient is free of symptoms, and (5) that for the first time pain may now be felt in the right hypochondrium and lead to tenderness on gentle deep palpation, and so to suspicion of the true diagnosis. This sequence is important and is similar to that pointed out many years ago by J. B. Murphy in cases of acute appendicitis.

I do not know whether the nocturnal bouts of sweating and rise in pulse rate had any connexion with the presence of gallstones, but it may be of significance that since the operation a year ago I have not had a similar attack. Finally, I learned the truth of the saying that "one is never too old to learn."

CHOLANGITIS

It may be difficult or impossible to distinguish cholangitis secondary to a common duct stone from acute cholecystitis. In general, the patient with cholangitis tends to have a greater constitutional reaction, with higher fever and often rigors but with lesser degrees of local tenderness than the individual with acute cholecystitis. The degree of hyperbilirubinemia tends to be lower in patients with acute cholecystitis.

INFLAMED OR LEAKING DUODENAL ULCER

In the case of a duodenal ulcer that is threatening to perforate or has even leaked slightly, the local findings may be similar to those of cholecystitis with local peritonitis, but a careful inquiry into the *history* will distinguish the two conditions. The pain that comes on about two and a half hours after meals and is relieved by taking food, the bringing up of "waterbrash" and acid eructations, the attacks of flatulence, and possibly the occurrence of melena may give a clear picture of ulcer. If time permits, the diagnosis of ulcer may be confirmed by X-ray or endoscopy.

FREE RETROPERITONEAL PERFORATION OF DUODENAL ULCER

Retroperitoneal perforation of a duodenal ulcer may be attended by severe collapse at the onset, but the condition quickly localizes and leads to tenderness and swelling in the right loin. The perinephric tissues become edematous, and there may be frequency of micturition and even hematuria from the irritation of the renal pelvis.

There is great pain on pressure at the erector-costal angle. The diagnosis is difficult, since a primary renal condition is likely to be suspected. It is, fortunately, a very rare condition.

BILIARY COLIC

Biliary colic (unassociated with inflammation of the gallbladder) is usually an antecedent of acute cholecystitis. The abdominal wall over the gallbladder is soft and yielding, though there may be local deep tenderness. The pain is usually radiating, being felt especially in the right subscapular area or in the midline over the lower thoracic spine. It may also be felt on the left side, in which case there may be a complaint of a sense of constriction around the waist. This feeling of constriction, when present, is very characteristic of biliary colic. A subnormal temperature is more common than fever.

RUPTURE OF THE GALLBLADDER

Although rupture of the gallbladder is not infrequent, it is quite uncommon for generalized peritonitis to occur because the perforation is usually very well confined in the form of a pericholecystic abscess. Such cases are indistinguishable from an ordinary acute cholecystitis, except that the attack by that time has usually been present and unrelenting for at least five or six days.

With free rupture of the gallbladder or of one of the bile ducts into the general peritoneal cavity, there are usually the history and symptoms suggestive of biliary colic or cholecystitis, with a gradual extension of the painful area downward until the whole abdomen is tender; distention of the intestines increases, and there is tenderness on rectal examination. Free fluid may sometimes be demonstrated. There is often a history of acute onset, with a subsequent remission of symptoms for a day or two and a final exacerbation of symptoms as the peritonitis spreads over the abdomen.

HEPATITIS

In hepatitis there is tenderness all over the liver, including the *lateral aspect,* as ascertained by pressure in the lower intercostal spaces laterally, as well as in the right hypochondrium. This sign serves for diagnosis except in those cases where hepatitis coexists with the cholecystitis. The onset of pain is rarely abrupt in hepatitis.

PLEURISY

In right basal pleuropneumonia or diaphragmatic pleurisy fever is usually higher (104°F or 105°F), there may be an initial rigor, the right hypochondriac tenderness is more superficial, by gradual coaxing the fingers may be pressed well into the subhepatic region, and there should be signs—at any rate, fine crepitations—at the base of the right lung and pleura. Pain on top of the right shoulder is much more likely to be met with in diaphragmatic pleurisy than in cholecystitis. The occurrence of hemoptysis in any doubtful case would suggest either pneumonia or a pulmonary infarct.

MYOCARDIAL INFARCTION OR ANGINA PECTORIS

One of the most frequent noncardiac conditions with which a patient is admitted to modern coronary care units is biliary colic or early acute cholecystitis. The midline position of some cases of biliary pain, especially when located high in the epigastrium, occasionally makes the distinction very difficult. If a true myocardial infarction evolves, the electrocardiogram and serum enzyme changes, when followed serially, serve to distinguish the conditions. We have on occasion been unable to distinguish repeated attacks of angina pectoris, particularly when they occurred postprandially, from episodes of biliary colic. Even a stress electrocardiogram may fail to make this distinction. Suffice it to say that during myocardial infarction the abdominal examination is usually normal but an electrocardiogram must always be taken in a patient with suspected cholecystitis.

11. Acute abdominal lesions arising in the left hypochondrium

The lesions that cause pain in this region are rather few in number, and thus the diagnosis usually can be made reasonably accurately. The left hypochondrium contains the fundus of the stomach, the spleen and its blood vessels, the tail of the pancreas, the splenic flexure of the colon, and, behind the peritoneum, the upper pole of the left kidney. The upper part of the abdominal aorta lies just to the left of the midline in contact with the body of the first lumbar vertebra. Each of these viscera may on occasion cause acute abdominal symptoms.

Acute pancreatitis. Acute pancreatitis is one of the most common causes of pain in the left hypochondrium, often occurring without any epigastric component at all. This is not to say that this presentation is the most common one in pancreatitis, for it is not. When considering left hypochondriac pain alone, however, pancreatitis is an extremely important cause. The pain may radiate straight through to the back or toward the tip of the left scapula. Vomiting and retching are frequent but produce little return.

Stomach. Gastric ulcers are rare in the fundus, and perforation of an ulcer is seldom found there. Sometimes a perforation may be found near the cardia.

Spleen. Rupture of the spleen is frequently caused by injury. The symptoms are described elsewhere in this volume.

Spontaneous rupture of a normal spleen is very rare, but an enlarged, diseased spleen may rupture and cause rapid and sometimes fatal hemorrhage.

Splenic infarcts may cause pain aggravated by breathing, and occasionally a friction rub may be heard over the infarct. These infarcts are common in sickle-cell crises (disease and trait) and occasionally result from lodgment of septic emboli in the spleen.

Aneurysm of the splenic artery. Pain is usually unassociated with other abdominal symptoms unless rupture with severe intraperitoneal hemorrhage occurs. Rupture tends to have a predilection for the pregnant female. The pulsatile mass is occasionally palpable in a slender patient, and splenomegaly is usually present. A plain X-ray may show an "eggshell" type of ringlike calcification, and in many instances several such rings can be seen since these lesions are often multiple. Arteriography may be necessary for final confirmation.

Aneurysm of the abdominal aorta. A small aneurysm may be extremely difficult to diagnose. I recall one patient whose only symptom was a constant pain along the peripheral course of the twelfth thoracic nerve. He was thoroughly examined by a physician, a neurologist, an orthopedic surgeon, and a radiologist, without anything pathological being found. The patient then went to a well-known clinic, where a similar negative result was obtained. Suddenly the patient died. At autopsy the cause of the disease and its only symptom was found to be a small aneurysm of the aorta that had burst into the intestine. Today, ultrasonic or CT examinations might well disclose such a lesion (Plate 5).

Carcinoma or stricture of the splenic flexure of the colon. This may cause severe pain that can usually be localized. The colon is here in contact with the sensitive lateral abdominal wall, and this localizes the pain. Constipation is a common symptom, and intestinal colic may be severe. Diagnosis can usually be made by radiography or colonoscopy. If several diverticula filled with the opaque substance are seen the symptoms are likely to be due to diverticulitis.

ACUTE PERINEPHRITIS

A discussion of right-sided perinephritis will be found in Chapter 7, but left-sided perinephritis is a more dangerous condition. The increased danger on the left side is chiefly due to anatomical reasons. The right kidney is slightly lower than the left, and on the right side the massive right lobe of the liver fills the right hypochondrium and prevents extension upward, while on the left side the soft, mobile fundus of the stomach and the soft spleen do not hinder any acute inflammation from extension upward toward the upper part of the diaphragm. Therein lies the danger.

Acute perinephritis is not commonly due to extension of a pyonephrosis; it is somewhat more frequently caused by metastasis of virulent septic bacteria from a focus elsewhere in the body (e.g., a boil or carbuncle). The germ usually settles in the cortex of the kidney and soon breaks through into the extensive retroperitoneal space. Suppuration occurs, and there is always the danger that infection may penetrate the diaphragm and cause pleurisy and empyema.

Signs and symptoms. Irregular fever will be noted, and there will be tenderness on deep pressure in the left subcostal region. Respiration may be painful and tenderness may occur on deep pressure in the left hypochondrium. If the diaphragm becomes involved, there may be tenderness over the lower left chest posteriorly and an X-ray will be necessary.

Before the advent of antibiotics this condition was very dangerous, but nowadays it should be prevented, though an incision may be needed to drain an abscess.

Plates

PLATE 1

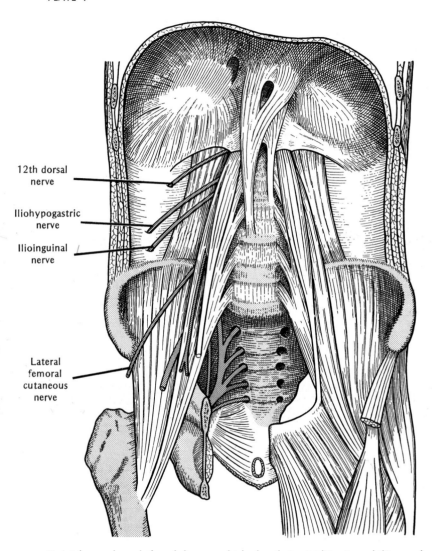

12th dorsal
nerve

Iliohypogastric
nerve

Ilioinguinal
nerve

Lateral
femoral
cutaneous
nerve

Parietal muscles of the abdomen which, by their rigidity, immobility, and tenderness, give important help in diagnosis of the acute abdomen. On the right side the twelfth dorsal nerve, iliohypogastric, ilioinguinal, and lateral femoral cutaneous nerves, are indicated.

PLATE 2

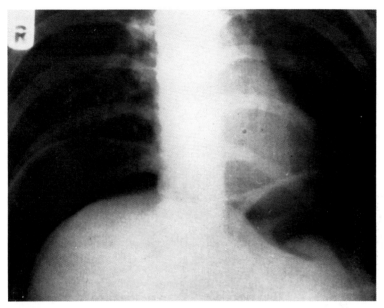

Showing large amount of free gas between the diaphragm and the liver in a case of perforated peptic ulcer.

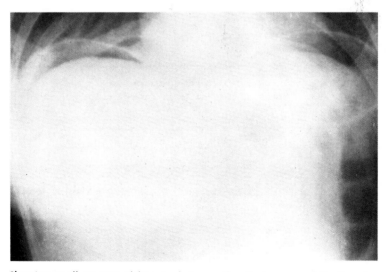

Showing small amount of free gas between the diaphragm and the liver in a case of perforated peptic ulcer.

PLATE 5

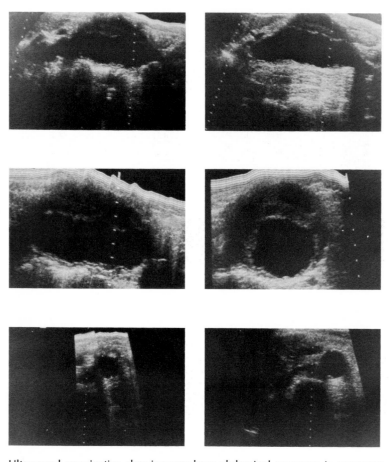

Ultrasound examination showing very large abdominal aneurysm in transverse section.

PLATE 6

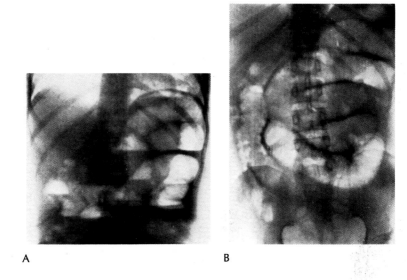

A   B

Ileal obstruction by adhesions nine days after repair of right inguinal hernia.
(A) Patient erect. (B) Patient supine. (C) Ileal obstruction by adhesions in right
iliac fossa, showing the value of the lateral radiograph with the patient supine;
horizontal X-ray beam.

C

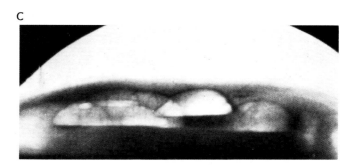

PLATE 13

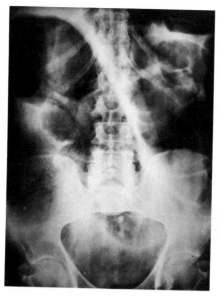

Λ

Obstruction of the colon by a carcinoma of the sigmoid.

(A) Patient supine. (B) Patient erect.

Massive distension of the large bowel from the cecum to the sigmoid where the obstruction was situated. In the erect film, colonic air fluid levels can be seen.

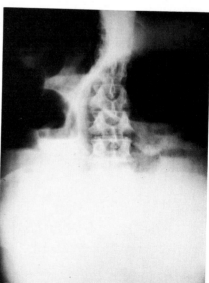

B

PLATE 6

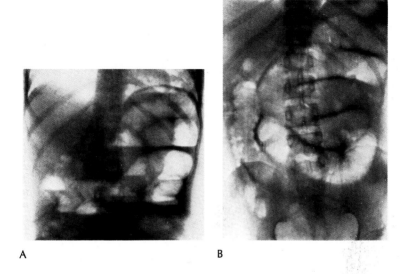

A B

Ileal obstruction by adhesions nine days after repair of right inguinal hernia. (A) Patient erect. (B) Patient supine. (C) Ileal obstruction by adhesions in right iliac fossa, showing the value of the lateral radiograph with the patient supine; horizontal X-ray beam.

C

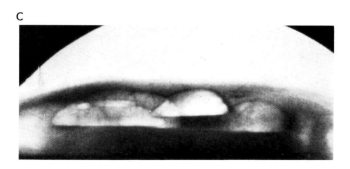

PLATE 7

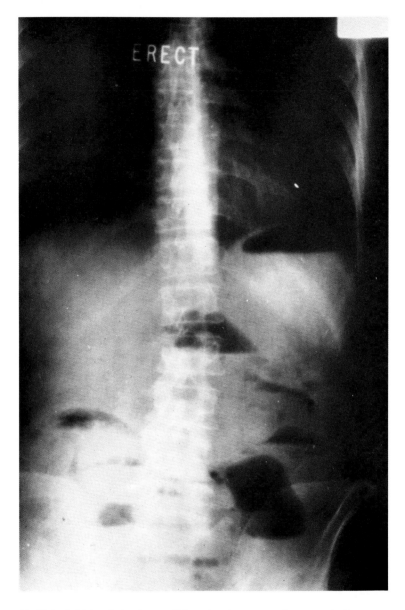

Multiple fluid levels in the small intestine in a case of intestinal obstruction.

PLATE 8

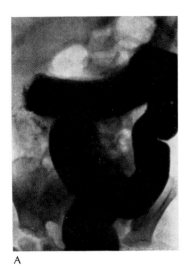

A

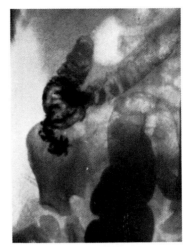

B

Series of radiographs taken during the administration of a barium enema in a case of intussusception.

(A) The opague barium stopped its advance in the transverse colon at the site of the intussusception.

(B) The enema is reducing the intussusception and filling the ascending colon. (C) The intussusception has been forced back to the cecum, which is filling with barium.

(D) There still remains a small part of the cecum which does not fill with barium. Operation showed this was due to the last unreduced part of the intussusception.

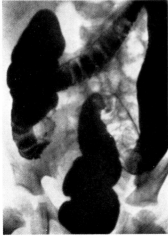

C

D

PLATE 9

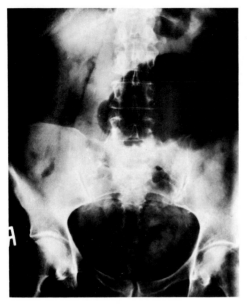

A

Volvulus of the cecum

(A) Patient supine. (B) Patient erect.

Note that the distended cecum orients itself away from the right iliac fossa.

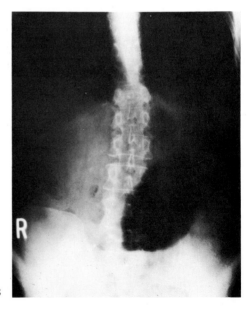

B

PLATE 10

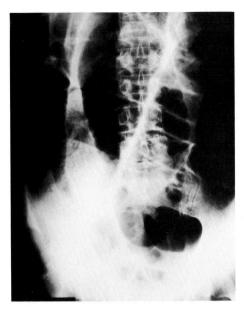

A

Volvulus of the sigmoid

(A) Patient supine. (B) Patient erect.

Note that the dilated sigmoid orients itself away from the left iliac fossa and towards the upper abdomen. The colon proximal to the closed loop of sigmoid is also dilated with gas.

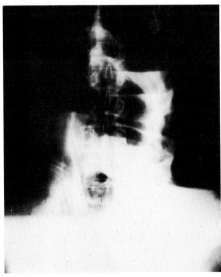

B

PLATE 11

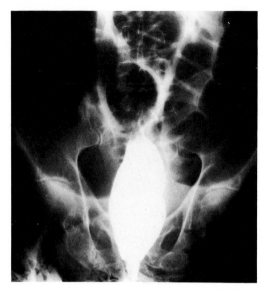

A

Barium enema of patient shown in Plate 10. Note the smooth tapered end of the barium column, the so-called bird-beak deformity characteristic of this condition.

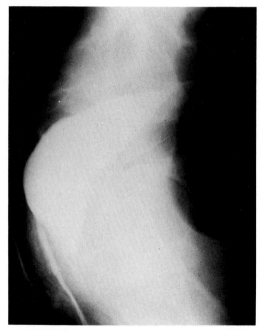

B

PLATE 12

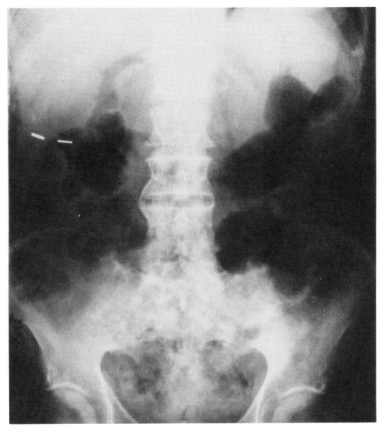

Plain film showing free air in biliary tree in patient with gallstone obstruction of ileum.

PLATE 13

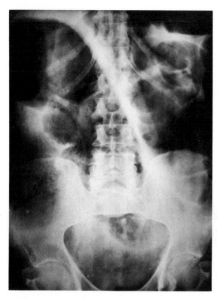

A

Obstruction of the colon by a carcinoma of the sigmoid.

(A) Patient supine. (B) Patient erect.

Massive distension of the large bowel from the cecum to the sigmoid where the obstruction was situated. In the erect film, colonic air fluid levels can be seen.

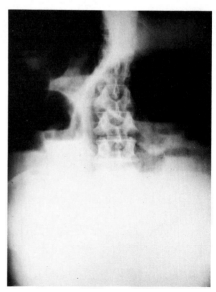

B

PLATE 14

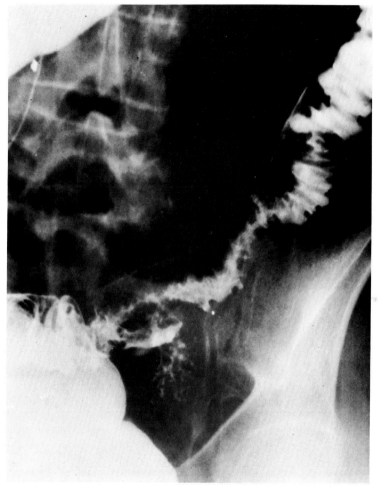

Barium enema showing a pericolic abscess and fistulous tract arising from sigmoid diverticulitis.

JEJUNAL DIVERTICULITIS

In recent years, many observations have been made on patients suffering from acute abdominal symptoms due to inflammation in one or more jejunal diverticula. When these diverticula are not inflamed there may be no acute symptoms, but the patient often complains of epigastric pain radiating down the left side of the abdomen and coming on at a varying time after the taking of food. Flatulence, borborygmi, and diarrhea may result from bacterial overgrowth in these "blind loops."

When acute abdominal symptoms supervene in a patient who gives the characteristic history, they are usually due to inflammation or rupture of a diverticulum. Pain in the upper part and left side of the abdomen, vomiting, local rigidity, and tenderness are the indications of a local peritonitis. If the leakage is gradual and a local abscess forms, one may feel a mass, and sometimes a partial obstruction of the small bowel may result. If, as sometimes may happen, the symptoms are more noticeable on the right side of the abdomen, the simulation of appendicitis may be almost complete.

12. The colics

True abdominal colic is always caused by the violent peristaltic contraction of one of the involuntary muscular tubes, whose normal peristalsis is quite painless. The violence of the contraction is usually produced in an effort to overcome some obstacle that prevents the passage of the normal excretion or secretion. The pain is due to the stretching or distention of the tube, and the agony produced may be as severe as any to which a human being can be subjected. The involuntary muscular tubes that may thus cause colic are:

1. The stomach and intestines.
2. The cystic, hepatic, and common bile ducts and the gallbladder.
3. The pancreatic duct.
4. The ureters.
5. The uterus and fallopian tubes.

The main feature of severe colic is the occurrence of acute, agonizing, *spasmodic* pain, which causes the patient to double up. It is associated in varying degrees with the symptoms consequent to excessive stimulation of the sympathetic nervous system, for example, pallor or lividity, weak pulse, vomiting, subnormal temperature, or coldness of the body surface. The pain is referred partly to the local site of origin and partly to the area of nerve distribution of the segment of the spinal cord with which the affected part is associated. The abdominal muscles, in common with many of the other voluntary muscles, may remain contracted and rigid during the height of the pain, but they relax when the spasm of pain subsides, and the fingers may then be pressed fairly easily into the abdominal cavity, though there may be local tenderness over the affected viscus.

In the general diagnosis of a colic the following points may help:

1. The pain usually comes in paroxysms lasting a variable time and often reflected by intermittent facial grimacing.
2. In colic the patient is usually very restless and flings himself about as if to find some relief from the pain that grips him. A flexed position of the body may be adopted during the pain.
3. Though occasionally rigid during the paroxysms, the abdominal wall is soft between the bouts of pain. In sudden acute peritonitis it remains rigid all the time.
4. In colic, pressure to the abdomen sometimes relieves the pain— an occurrence not usual in other acute conditions.
5. In many of the colics the distribution of pain is almost diagnostic (Fig. 23).

INTESTINAL COLIC

Colic of the small intestine is sometimes caused by enteritis. It is more severe in cases of organic obstruction to the small intestine. The pain is acute and gripping, is referred to the epigastric or umbilical region, and is accompanied sometimes by local areas of distention where gurgling sounds may be heard and sometimes felt by the palpating hand. Vomiting may take place, and peristalsis may occasionally be seen through the abdominal wall or borborygmi heard on auscultation. When due to enteritis, the pain is generally soon followed by diarrhea, the nature of which may give the clue to the cause of the pain. When due to organic obstruction, the attacks will occur from time to time until a final attack of acute intestinal obstruction occurs. When severe pain, assumed to be

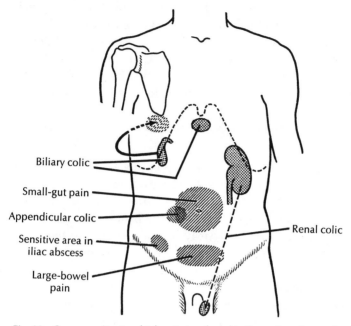

Fig. 23. Common sites to which pain is referred in the various forms of colic. (See also Fig. 2.)

due to intestinal colic, persists for more than three or four hours, the condition is generally one needing surgical intervention.

Lead colic is a form of colic of the small intestine that is accompanied by constipation. The pain may be extreme and the collapse of the patient severe. During the spasms of pain the abdominal wall may be rigid. Other signs and symptoms pointing to lead poisoning (blue line on the gums, severe constipation, local paralysis, etc.) may be present.

Colic of the large intestine is very common, but the pain occasioned by the large bowel is seldom so acute or prostrating as that resulting from colic of the small bowel. The pain is referred chiefly to the hypogastrium. The causes are either severe constipation due to hard scybalous masses, colitis, or dysentery or some form of stricture of the large bowel, usually carcinomatous. Pain in the colon is often more accurately localized by the patient than is the pain of small-bowel colic.

Diagnosis. In any case of intestinal colic diagnosis is greatly aided by the history, for the history of eating tainted food or the knowledge that similar attacks have occurred in other members of the household, the occurrence of previous similar attacks in one who is a painter by trade, or an account given of former bouts of dysentery, may throw considerable light on the problem. By obtaining a careful history it should also be possible to exclude appendicular colic. The occurrence of diarrhea would usually exclude intestinal obstruction, but it must be recollected that loose motions are sometimes seen in cases of PARTIAL obstruction or with an inflamed pelvic appendix. Local or diffuse peritonitis must be excluded by noting the absence of rigidity when the severe pain subsides, the relief obtained by gentle pressure on the abdomen, and the absence of tenderness of the pelvic peritoneum.

If attacks of small-intestine colic recur from time to time and lead to loss of weight, the possibility of organic obstruction (by adhesions, bands, tuberculous stricture, or neoplasm) must be considered, nor is it wise to delay too long before advising radiographic examination of the intestinal tract.

Similarly when there are recurrent attacks of distention of the large bowel, accompanied by colic and constipation, it is wise to suspect carcinoma or stricture of the colon or rectum, and certainty on the question must be obtained by the proper diagnostic procedure (see Chapter 13).

BILIARY "COLIC"

Biliary "colic" may be caused by a small gallstone attempting to pass through the cystic, hepatic, or common bile ducts. Because these structures have the same segmental innervation, *it is often impossible to distinguish the pain of a stone in the cystic duct from pain caused by one in the common bile duct,* except by the subsequent course of events (see Chapter 11). *The term biliary "colic" is es-*

sentially a misnomer since the pain arising from passage of a biliary stone is steady and not paroxysmal. The pain results from distention of the biliary tree, and because the muscle wall of the gallbladder and especially of the bile ducts is sparse, strong intermittent contractions usually do not occur. *Failure to take note of the almost always steady nature of biliary "colic" will lead to many faulty diagnoses.* The most common place in which gallstones form is the gallbladder; biliary "colic" is usually caused by a stone passing through the cystic duct. Large stones are too big to enter the narrow canal of the cystic duct; if they cause symptoms, it is by gradually ulcerating either into the first or second part of the duodenum or into the hepatic flexure of the colon, thereby causing acute or subacute obstruction of the intestine. Stones in the common bile duct more commonly lead to obstructive jaundice.

It has been commonly observed that biliary "colic" occurs more often at night than during the day. This may be partly due to the anatomical position of the gallbladder, which, in the upright position of the body, lies vertically, with the opening of the cystic duct above and the fundus of the viscus lower down. Thus, during the daytime any stones would fall to the bottom or fundus, while during the night when the patient is in bed, the gallbladder would be horizontal and the stones, or at least some of them, might approach nearer to the opening of the cystic duct.

When gallstones are present in the gallbladder, the eating of an evening meal containing a good deal of fatty food may be followed by an attack of biliary "colic" the following night. Such a meal would cause a plentiful secretion of bile that would also fill the gallbladder with fresh bile; as digestion proceeds, the gallbladder would begin to empty and the flow of bile may help to move a small stone into the cystic duct, with resulting biliary "colic."

When once a stone has entered the cystic duct, it may proceed until it reaches the common bile duct or even the duodenum, when the "colic" will cease. If, however, it fails to reach the duodenum but stops in the cystic or common duct, then attacks of "colic" may occur either by day or by night.

Pain varies considerably in intensity according to the difficulty the stone experiences in traversing the ducts, and is usually sudden in onset and severe in intensity. Vomiting is common and collapse may be so severe that the observer may consider the patient *in extremis*. Generally the patient writhes in agony, but in the worst cases may lie still, with a pinched, pale face, cold extremities, and weak pulse. The pain is felt:

1. In the right hypochondrium, which is usually tender on pressure.
2. Very often at the midline of the epigastrium in a sharply confined way.
3. Below the inferior angle of the right scapula.
4. Sometimes radiating directly through to the lower thoracic spine.
5. Occasionally at the same levels on the other side of the body.
6. Infrequently it may be referred to the right acromial and lower cervical regions.

The common position of the pain corresponds to the distribution of the nerves from the eighth and ninth dorsal segments of the spinal cord.

Diagnosis. Diagnosis is usually fairly clear on account of the intensity and distribution of the pain and the absence of local abdominal rigidity when the pain passes off. *When rigidity persists* in the right hypochondrium, there must be accompanying cholecystitis or *local peritonitis.* A leaking or inflamed duodenal ulcer also causes persistent rigidity, but in the case of ulcer there should be a history suggestive of that condition (see Chapter 8). The abdominal pain of acute porphyria has been mistaken for biliary "colic."

RENAL COLIC

Renal colic is caused by the passage of a small stone, a portion of blood clot or inspissated pus, oxalate crystals, or uratic debris down the ureter, by the impaction of a stone in the renal pelvis, or rarely by sudden kinking of the ureteropelvic junction when there is an unduly mobile kidney. It is sometimes, but by no means always, accompanied or followed by hematuria.

The symptoms are usually characteristic. The patient is seized with sudden pain starting in the loin and often radiating to the corresponding testicle or groin or (in women) to the vulva. In some cases there is extensive superficial hyperesthesia of the abdominal wall, either in front or posteriorly. In severe cases there is violent restlessness. Vomiting is common. There may be frequency of urination and pain on performing the act. Hematuria may accompany or follow the pain. Renal colic in males by no means always radiates to the testicle, and it must be remembered that pain due to inflammation of the appendix is sometimes felt in the testicle.

In some cases, the pain is steady and not paroxysmal. When this happens, the distribution of the pain will help make the diagnosis.

Diagnosis. The distribution of the pain is characteristic, and local examination of the loin may reveal a tender and possibly enlarged kidney. An X-ray examination may show a stone to be present, but there are many cases of renal colic in which no stone is seen by plain X-rays, yet a small stone is passed later. Minor degrees of ureteric colic are frequently misdiagnosed as appendicular colic, especially when a plain radiograph is unsuccessful in revealing the small calculus. Intravenous pyelography is usually diagnostic.

UTERINE COLIC DYSMENORRHEA

The pain caused by the uterus in its attempt to expel either a fetus, a polypus, a membranous cast, or even a blood clot may be very severe. The pain is referred chiefly to the lower lumbar region, but in severe cases it may radiate down the thighs and over the hips. Vomiting and retching may occur.

Spasmodic dysmenorrhea may cause the patient to writhe. Since with any abdominal pain in women the menstrual history would be inquired into, and with spasmodic dysmenorrhea vaginal and pelvic examinations would show no special reason for pain of uterine origin (apart perhaps from a pinhole os and a small cervix), there should be little difficulty in diagnosis.

GASTRIC COLIC

Attacks of severe epigastric pain of a colicky nature sometimes occur in those who suffer from pyloric stenosis due to ulcer or neoplasm, but this condition is surprisingly rare. In fact, *severe gastric outlet obstruction ordinarily produces no pain whatsoever (though the causative ulcer may)*. Perhaps gastric colic is uncommon because the onset of obstruction is usually insidious and decompensation of the antral musculature gradually supervenes. In such cases the peristaltic wave can often be seen going from the patient's left to right side, and the outline of the dilated stomach may readily be recognized through the abdominal wall. The pain is referred to the area of distribution of the tenth dorsal segment, chiefly on the left side.

Severe pain may occasionally be caused by a sudden hemorrhage taking place into the stomach. In such cases there is usually a history suggestive of gastric ulcer. The severe pain, collapse, and epigastric tenderness may lead one to diagnose intestinal obstruction or gastric perforation. But the sudden anemia, the vomiting of altered blood (which usually but not always occurs), and the lax abdominal wall give the clue to the correct diagnosis.

PANCREATIC "COLIC"

Obstruction to the duct of Wirsung may give rise to severe pain, which sometimes radiates to the left shoulder. The pain is almost always steady rather than paroxysmal and is worsened by the eating of food. Thus, as in the case of biliary "colic," the term pancreatic "colic" is misleading. The pain results from distention of the pancreatic ducts, which have little musculature. Vigorous contractions of the bile ducts or the pancreatic ducts has never been observed either radiologically or at operation, in contrast to those of the ureter or small intestine.

13. Acute intestinal obstruction

Acute obstruction of the intestine in the form of strangulated hernia was one of the first abdominal emergencies to be referred to the surgeon for treatment, while obstruction accompanied by similar symptoms, but due to internal causes that were not so obvious as an external hernial swelling, was among the latest urgent abdominal cases to be referred by the physician to the surgeon. If there is any condition in which early diagnosis and operative treatment and avoidance of attempts at purgation are necessary, it is intestinal obstruction.

The pathology and causation of acute intestinal obstruction are questions far too big to discuss fully in a small book. We are here concerned only with the common causes and the main types of cases which come for diagnosis. It is not always essential in diagnosis to know the exact cause of the obstruction, though every effort should be make to ascertain it as accurately as possible. It is useful to have a knowledge of the proportion of cases due to the main pathological causes of obstruction. The following list is a compilation of the overall causes of intestinal obstruction (both small and large bowel) taken from thirteen reported series comprising a total of 12,731 patients. The differences in the distribution of etiologies between adults and children are readily apparent. These findings do *not* apply to tropical regions.

Adults		Children	
Hernia	41%	Hernia	38%
Adhesions	29%	Pyloric stenosis	15%
Intussusception	12%	Ileocecal intussusception	15%
Cancer	10%	Atresias & annular pancreas	14%
Volvulus	4%	Adhesions	7%
Miscellaneous	4%	Miscellaneous	11%

When the bowel is completely obstructed, the course of the disease is inevitably fatal unless the obstruction is relieved by:

1. The spontaneous rectification of the condition—which is very rare.
2. Formation of an external fecal fistula.
3. Operative interference.
4. The passing of a long intestinal tube.

The third method is nearly always advisable. A long intestinal tube is useful and safe only under certain circumstances. So long as the obstruction is *complete,* exclusion of strangulation is virtually impossible (see below). Therefore, intestinal intubation is advisable only in cases of *partial* small bowel obstruction, especially in the early postoperative period or after a recent bout of peritonitis. Until comparatively recently, the mortality after operation in cases of intestinal obstruction (excluding strangulated external hernias) was high, but during the last two decades better results have been obtained. The main desideratum is to diagnose the case early. There are few cases of intestinal obstruction that cannot be remedied or alleviated if brought to the surgeon within twelve hours of the onset of symptoms.

Intestinal obstruction may exist in a chronic or subacute form for a considerable period before an acute attack ensues. In the chronic form, the symptoms are similar in kind but different in degree from those resulting from an acute attack. Chronic obstruction, if uncorrected, sooner or later terminates in an acute attack.

Symptoms

Acute intestinal obstruction is always to be regarded seriously, but at the outset two points must be emphasized: first, the variability of the symptoms according to the site and cause of the obstruction, and second, the deceptive nature of those symptoms unless great care is taken by the observer. It is generally true that the higher up in the gut is the obstruction, the more severe are the symptoms. It is indeed usually possible to estimate which part of the gut is obstructed by considering the symptoms, and cases may be roughly divided into three groups according to whether the obstruction is in (1) the upper small intestine, (2) the lower small intestine, or (3) the large bowel. Let us first consider:

General symptoms and signs of acute intestinal obstruction

Pain	Distention
Vomiting	Tenderness of abdomen
Constipation (inability to pass feces	Visible peristalsis
or flatus)	Shock

These symptoms and signs are seldom if ever all present at the same time, and the value of each symptom must first be discussed.

Pain. Pain is often very severe from the outset. It is referred to the epigastric and umbilical regions, or sometimes to the hypogastrium, and usually comes on in bouts or spasms, though if a considerable segment of mesentery is implicated or if peritonitis has supervened, the pain may be continuous. The spasmodic pain is due to the peristalsis of the intestine trying to overcome the obstruction. This can easily be demonstrated during the examination of an intussusception, when the soft tumor may be felt to harden just before the screaming of the infant. The higher up in the intestine it is located, the more severe the pain tends to be.

Vomiting. Vomiting is an almost constant feature but varies greatly in the different forms. The higher up in the intestine the obstruction occurs, the sooner vomiting sets in and the more violent is the regurgitation. *In obstruction of the large bowel, vomiting may be absent* but nausea and anorexia are usual. In obstructive vomiting, first the stomach contents are expelled, then green bilious material appears and, if the obstruction is some way down the small intestine, the color of the vomit changes to yellow or greenish-brown and becomes feculent. A feculent vomit, in the absence of peritonitis, is diagnostic of intestinal obstruction, though it should be regarded as a late symptom of that condition. Feculent vomit almost always signifies obstruction low in the small intestine; it is occasionally associated with colonic obstruction, but only if the ileocecal valve is incompetent. The nature of the vomit is therefore to be watched very carefully, and the vomited material should never be thrown away until seen by the medical attendant. True fecal vomit is rare and is only seen when a communication exists between the colon and stomach, and rarely even then.

Constipation. Constipation and the nonpassage of flatus are symptoms of intestinal obstruction but are not always evident at first. If the bowel is occluded at any spot, it is clear that no contents can pass the occluded area, but the gut below the stoppage can empty itself, so that for a time motions may be passed. As soon as the lower bowel is emptied (either naturally or by enemas), neither flatus nor feces pass. There are many acute cases in which constipation should be regarded as a sequel rather than a symptom, for valuable time may be lost in the attempt either to open the bowels or to prove that the obstruction exists. *If the other symptoms of intestinal obstruction are present, it is unwise to wait twelve or twenty-four hours to demonstrate constipation.* It is advisable to point out that in acute intussusception there is occasionally incomplete obstruction, so that some brown or yellowish fecal material may be passed per anum in addition to the blood and mucus.

Distention. Distention is usually late in appearing in the acute cases or in those in which the upper part of the small intestine is affected. But always, sooner or later, it is in evidence. In obstruction of the large bowel and the lower end of the small bowel, it may be apparent by the time the symptoms become acute enough to call the serious attention of the patient. It is indeed because of the relative slightness of the pain and vomiting that the distention in these cases is allowed to proceed so far.

Distention at first may be merely local, due to the dilatation of the coils of gut immediately above the obstruction. In some cases, where ordinary clinical examination does not show any definite distention, a radiograph may reveal local distention of the intestine. In obstruction of the end of the ileum a local hypogastric distention may first be observed, and in volvulus of the sigmoid the outline of the affected coil of large bowel may stand out very distinctly. In subacute or partial obstruction of the lower end of the small bowel the distention gradually affects coil after coil, so that when the patient comes under observation with acute symptoms the typical *ladder-pattern type of distention* is seen on looking at the abdomen (Fig. 24).

Tenderness. Tenderness of the abdomen is not usually found until distention appears. Pressure over a distended coil is generally painful. A completely strangulated coil of small intestine quickly becomes distended, tense, and tender on pressure. In the later stages of obstruction, when peritonitis has ensued, there may be general pain all over the abdomen. *Rigidity* of the abdominal wall is unusual except in those cases where there is some local peritonitis around an obstructed or strangulated section of bowel. In obstruction of the small intestine, a mass may be appreciated if a closed, strangulated loop is present.

Peristalsis. Visible peristalsis is not a constant accompaniment of obstruction but is diagnostic when present, except in some few very thin persons in whom the normal peristaltic movements of the intestines can be seen through the abdominal wall. It is not usually

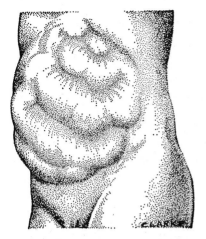

Fig. 24. The ladder pattern of abdominal distention (indicating obstruction of the lower ileum).

seen in the very acute cases of strangulation, but in subacute obstruction it is more frequently noted and is very valuable in diagnosis. In some instances peristalsis may be accompanied by a gurgling of gas, which may occasionally be heard to pass through a narrowed part of the gut. If a stethoscope is applied to the abdomen during a bout of pain, it may be possible to hear sounds caused by fluid passing through the site of obstruction under the pressure caused by the peristaltic wave. However, when distention is great or has been present for some time, the abdomen may be very quiet and without borborygmi. Although high-pitched, tinkling rushes are helpful when the remainder of the clinical picture is consistent with intestinal obstruction, their absence should not stay the hand of the surgeon.

Shock. In severe cases the pulse may be weak, the skin cold and sweating, the temperature subnormal, and the pupils dilated, but as a rule the symptoms are not so severe as this. True shock is a very late finding and means that the diagnosis has been missed for too long.

SYMPTOMS DUE TO DIFFERENT TYPES OF OBSTRUCTION

The features of an attack of obstruction vary according to (1) the part of gut obstructed, (2) whether or not the mesentery with the contained blood vessels is also affected, and (3) the completeness or incompleteness of the obstruction.

The part of the gut obstructed. The symptoms due to obstruction—(1) high up in the small intestine, (2) low down in the small intestine, and (3) in the large bowel—can be roughly differentiated.

1. *Obstruction high up in the small intestine.* This leads to acute symptoms, vomiting comes on very early and is frequent and violent, initial pain is greater, and distention is not an early feature. The vomit is green and bilious. Such symptoms are typically seen when a large gallstone ulcerates into the duodenum. Obstruction of the duodenum by a cicatrized ulcer may sometimes be acute owing to sudden spasm and edema around the ulcer. In such cases everything taken by mouth is returned, but no feculent vomit occurs and sometimes peristalsis of the stomach may be seen. The same symptoms are seen in infants suffering from congenital hypertrophic pyloric stenosis. Distention is seen only in the epigastric region. With obstruction of the upper jejunum the symptoms are still very acute and the vomiting begins early and is frequently repeated, while distention is not at first a noticeable feature unless there is a strangulation of a large coil of gut. The farther down the jejunum the obstruction, the less acute the symptoms.

2. *Obstruction of the lower part of the small intestine.* The symptoms are less severe than those just summarized. Shock and pain may be present, but vomiting is a little later in onset and some time elapses before feculent vomiting occurs. Distention comes on after a few hours. In subacute cases the ladder pattern of distention is seen, and peristalsis is often visible. When there is obstruction in the lower ileum, considerable fluid is still absorbed from the bowel, the magnitude of volume deficits may not be great early, and the

secretion of urine is only slightly diminished. With upper jejunal obstruction, volume deficits are large, even early after onset, and there is a great diminution in the amount of urine secreted.

3. *Large-bowel obstruction* (Fig. 25). Pain is much less acute, shock is comparatively insignificant (except in some cases of volvulus and intussusception), vomiting is a fairly late and infrequent symptom, while distention almost from the onset of the acute attack is the rule. An exception must be made in the case of intussusception, for in these cases distention is not an early symptom and should not be awaited, since a distended abdomen accompanying an intussusception generally means that strangulation has supervened.

The consequences of large-bowel obstruction will be influenced by the competence of the ileocecal valve. If the valve is incompetent (as it is in about 15–20 percent of cases of colonic obstruction) and allows reflux into the small intestine, the patient's condition and the radiological picture often resemble those found in partial obstruction of the small intestine. When the ileocecal valve is competent, however, a closed loop exists between the obstructing mechanism and the valve, and gangrene of the cecum as a result of severe distention is a constant threat. Tenderness in the right iliac fossa in such a case usually suggests that this complication is imminent or has already occurred.

Compression of the blood vessels. Strangulation of bowel. When the vessels in the mesentery of a coil of gut are compressed, first the veins and later the arteries become occluded. There is, in fact, a strangulation of the coil, and the gut soon becomes gangrenous. This *strangulation of the coil of gut adds greatly to the immediate danger*, for the obstruction is *complete* and, unless the condition is soon relieved, perforation of the bowel will occur and cause a fatal general peritonitis. This condition is commonly brought about by bands, by external or internal hernia, or by volvulus. The onset of symptoms is usually sudden, accompanied by great pain and early vomiting. Because intense symptoms usually begin early and bring

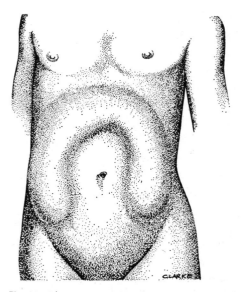

Fig. 25. The appearance of moderate distention of the large gut.

the patient to the hospital soon, distention is often absent until late in the course of the illness. Collapse and severe shock are late manifestations. If the gut is completely within the abdomen the local peritonitis will tend to become general, and by the time the case comes to the surgeon *there may therefore be definite abdominal rigidity over the affected part of the abdomen.* A mass consisting of the closed loop is sometimes felt.

The extent or completeness of the obstruction. The symptoms of obstruction vary greatly with the degree of occlusion of the lumen of the gut. When the obstruction is partial, or if the muscular contractions can to any extent force the contents past the obstructed part, the symptoms are less severe. Obstruction may result from a kinking of the gut due to adhesions between the bowel and the abdominal wall, another viscus, or a tumor. In these cases the onset is more insidious, the pain is not so severe and may have intervals of intermission, the vomiting may for a day or two be slight in degree, and distention comes on gradually. Quite commonly such cases do not come to the surgeon until feculent vomit has appeared. In these cases the abdominal wall is quite flaccid, since peritonitis is absent, for the gut remains intact, and its blood supply in the mesentery is unimpaired. This type is the variety that may occur after abdominal operations, when a certain amount of pain, vomiting, and distention is not unusual. There is therefore all the greater need to watch such cases carefully.

In the gradual narrowing of the gut lumen due to stricture, subacute volvulus, or chronic intussusception, and in cases where only a part of the lumen of the bowel is nipped in a hernial aperture (Richter's hernia), the onset of symptoms is still more gradual, vomiting is less, and the pain is intermittent.

A Richter's hernia of the cecum (of which one instance has occurred in the author's practice) may be unaccompanied by any obstruction of the bowel, and the herniated part of the gut wall may become gangrenous without giving rise to any acute symptoms.

When omentum only is strangulated there are subacute symptoms of obstruction, pain, vomiting, and some distention, but the vomiting never becomes feculent and the obstruction is never complete, for flatus and feces are brought away by enemas. The general symptoms in omental strangulation are usually slight, but if a large mass becomes gangrenous (e.g., an umbilical hernia) serious symptoms may result.

Diagnosis

DIAGNOSIS OF OBSTRUCTION OF THE SMALL INTESTINE

When a patient is seized with acute abdominal pain, becomes weak with a feeble pulse, cold extremities, an anxious look, and sweating skin, and soon begins to vomit first the stomach contents, then bile, then yellowish material that becomes brownish and feculent-smelling, while the abdomen remains flaccid, flat, and not tender, that patient is suffering from acute obstruction of the small intestine. The diagnosis should be made without waiting for distention to appear, nor—if the above symptoms are present—is there any need to demonstrate constipation. If the symptoms are not so acute there will be additional signs, for there is an inverse relation between the acuteness of the symptoms and the probability of the presence of signs. The higher up in the gut the obstruction, the more frequent the pains. Attacks of pain come on every three to five minutes in the upper ileum and every six to ten minutes in the lower ileum. When pain and vomiting and shock are slight, distention and visible peristalsis are more likely to be seen, and constipation, tested by two enemas, can usually be demonstrated. Thus it is that when acute supervenes upon chronic obstruction, both symptoms and signs are in evidence.

DIFFERENTIAL DIAGNOSIS OF ACUTE OBSTRUCTION
OF THE SMALL INTESTINE

Diagnosis must be considered either before distention has developed or after that sign has appeared. Always remember that almost half the cases of acute obstruction are due to strangulated hernias, and therefore *always examine first all the hernial orifices*, particularly both femoral rings.

Distention absent. When no distention is present, acute obstruction has to be distinguished from all the other acute abdominal catastrophes. The pain is often characteristic in type, occurring in bouts of acute intensity during which the severity is indicated by the patient's drawn features; it is usually central in position except in those cases in which a coil is strangulated, when the pain may be referred to the site of strangulation. Sometimes borborygmi may be heard passing along the bowel. The stethoscope should be applied to the abdomen and auscultation carefully carried out during the time that corresponds to a bout of pain. During the intervals between the bouts the patient's face shows apprehension of the return of the pain. From the acute inflammation—perforated gastric ulcer, pancreatitis, appendicitis with peritonitis, chole-cystitis—it is distinguished by the absence of rigidity and by the more frequent vomiting, *which tends to become feculent.* Renal and biliary colic are distinguished by the location and radiation of the pain and the absence of feculent vomit. Neither stoppage of the bowel contents nor distention follows the other colics. In torsion of a viscus (ovarian cyst, testicle) the vomiting does not become feculent. Gastric crisis is excluded by finding no other sign of tabes. It is usually sufficient to test the knee jerks and the pupillary reactions.

In many cases in which the diagnosis of obstruction of the small intestine is in question, great help may be obtained from a study of a plain X-ray of the abdomen. Normally little or no gas or fluid can be seen in the regions of the abdomen occupied by small gut. When there is obstruction, however, the contents are dammed back, and a plain radiograph will often show small collections of gas, each lying on top of a minute pool of liquid whose fluid level is charac-teristic. This aid is particularly helpful in the early stages of obstruction of the upper jejunum, when no distention may be observed. Two X-ray examinations should be made, one with the patient in the vertical position and the other with the patient lying in the horizontal position (Plates 6 and 7). If there is still a question during the attack, a contrast study *with barium* by mouth can often establish the diagnosis.

Volvulus of the stomach. In considering the diagnosis of obstruction in the upper part of the small gut, one must exclude volvulus of the stomach, a rare but well-recognized condition.

Acute volvulus of the stomach occurs chiefly in older people. It gives rise to violent and repeated vomiting, followed (after the stomach is empty) by continuous retching. The vomit contains no bile. The epigastrium becomes distended, but the lower abdomen remains empty and retracted. A stomach tube usually cannot be passed into the stomach. Symptoms of collapse ensue if the condi-tion is not remedied by operation. Gastric volvulus is commonly associated with large paraesophageal hiatal diaphragmatic hernias.

Distention present. When distention is present one must consider uremia, mesenteric thrombosis or embolism, or the late stage of peritonitis due to any cause.

Uremia should be diagnosed by careful consideration of the history, by examination of the urine for albumin and the blood for urea nitrogen, and maybe by detecting enlargement of the kidneys. Sometimes in uremic patients there may be a mere trace of albumin in the urine, but a low specific gravity would make one suspicious, and examination of the heart and determination of the blood pressure might throw light on the case.

Acute blockage of the mesenteric arteries or veins by an embolus or thrombus may lead to symptoms indistinguishable from those due to internal strangulation. In both cases the vascular disturbance leads rapidly to serious changes in the bowel. Sometimes a palpable mass is formed by the affected coil or gut. A history of recent endocarditis or the presence of auricular fibrillation might point to a possible embolism, while hepatic disease, the existence of thromboangiitis obliterans, or previous thrombotic trouble may sometimes suggest the possibility of mesenteric thrombosis. Though there is no mechanical obstruction in the intestines in a case of vascular occlusion, the affected gut soon becomes paralyzed and hemorrhagic, and the blood that is almost always poured into the bowel lumen as a consequence of the infarction sometimes passes into the gut farther on (see also Chapter 16).

Unless promptly dealt with, mesenteric thrombosis is soon followed by peritonitis.

DIAGNOSIS OF STRANGULATION

When the diagnosis of acute obstruction of the small bowel has been made with reasonable certainty, the next question to answer is whether there is strangulation of the gut, for nonoperative treatment by passing an intestinal tube and applying suction is not applicable when there is strangulation of the gut. A strangled coil usually soon becomes tender on pressure and, if near a sensitive part of the parietal peritoneum, may later lead to overlying muscular rigidity. When, however, the affected coil lies in the pelvis, these features are less in evidence and this differential diagnosis cannot be relied on. It is a truism that *so long as a mechanical small-bowel obstruction is complete, the presence or absence of strangulation cannot be excluded with certainty until the abdomen is opened.* Fever, leukocytosis, tenderness, rigidity, shock, and loss of peristaltic sounds are all late findings and, when present, usually indicate that delay has been excessive. Although X-rays may show a single distended coil in the shape of a coffee bean, distention is usually so late that the X-rays are interpreted as normal in about half of all cases of strangulation.

It may be impossible to distinguish between a late case of intestinal obstruction and late peritonitis unless the history gives some indication. In late peritonitis there is usually paralytic intestinal obstruction, while in the late stages of mechanical obstruction there is frequently peritonitis. In peritonitis, however, the vomit is seldom so definitely feculent as in late mechanical obstruction.

OBSTRUCTION OF THE SMALL INTESTINE BY A GALLSTONE

A gallstone that causes obstruction generally ulcerates into the duodenum, causing the symptoms of very acute obstruction, that is, severe pain and frequent vomiting. The vomit may contain blood from the ulcerated opening, and this may lead to an erroneous diagnosis of peptic ulcer. The stone passes on gradually and the symptoms abate considerably, but in a day or two the stone stops again at the lower end of the ileum (which is the narrowest part of the small intestine) and the symptoms recur. Therefore, if one is faced with a case involving symptoms of obstruction of the *upper* small gut in which these symptoms subside but are followed in a day or two by others suggestive of obstruction of the *lower* small gut, the cause is likely to be a gallstone (Fig. 26). This is the more likely if there has been a history suggestive of gallstones or if the patient is an obese woman of middle or advanced age. Occlusion of the small intestine by a gallstone is the only type of obstruction in which I

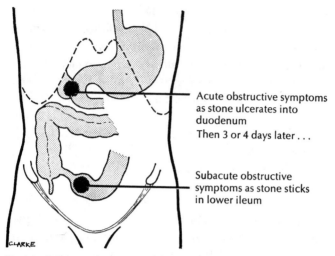

Acute obstructive symptoms
as stone ulcerates into
duodenum
Then 3 or 4 days later . . .

Subacute obstructive
symptoms as stone sticks
in lower ileum

Fig. 26. Gallstone obstruction of the small intestine.

have seen feculent vomiting come on very early and persist (with
diminishing frequency) without any serious collapse or abdominal
distention supervening. What little distention develops may be
concealed by a fat abdominal wall, so that the abdomen may actually appear normal. The character of the vomit remains feculent,
though there may be longer intervals between the bouts of pain and
vomiting. Considerable distention may develop toward the end of
the case.

The above characteristics make gallstone obstruction of the small
intestine more easy to diagnose than would be considered likely
from the rarity of the conditions. Some of these large gallstones
contain material opaque to X-rays, and a plain X-ray of the abdomen
will then show the stone in the intestine and guide one as to the best
place for the incision. If the radiograph should give evidence of air
in the biliary passages (Plate 12), the observer would be justified in
considering gallstone ileus as the probable cause of the symptoms.
In one patient under my care, the radiograph also showed other
large stones in the gallbladder, which in due course were passed
along the intestine and evacuated per anum.

BOLUSES OF INDIGESTIBLE FOOD

Sometimes the small intestine may be obstructed by a mass of indigestible food, for example, orange pith, dried fruits, cherry-stones, or (in America) persimmon, which causes pain, vomiting, and increasing distention, but as a rule the constitutional symptoms in these cases are not so severe. This complication is particularly likely to occur after gastrectomy, when the grinding function of the stomach is lost, and especially in edentulous patients. Even in individuals with an intact stomach, boluses of food may cause obstruction of the small intestine when a narrowed area is present, such as that caused by Crohn's disease of the ileum. I have been impressed by how frequently this is the obstructing mechanism in patients with Crohn's disease. It is usually but not always relieved by nonoperative treatment.

INTESTINAL PSEUDO-OBSTRUCTION

Recently, a new syndrome has been recognized consisting of recurrent attacks of what appears to be partial obstruction of the small intestine. Initially, there is no history of previous abdominal operation, and no groin or other hernia is demonstrable. Often other members of the patient's family have been victims of the same type of attack. Gastroesophageal reflux is common in these patients. The syndrome is caused either by defects of the circular musculature of the intestine or by abnormalities in autonomic innervation. Occasionally, it is a manifestation of progressive systemic sclerosis, in which case Raynaud's phenomenon is often an associated condition. Jejunal diverticula are likely to be found in those cases attributable to defects in the circular muscle. Because of the picture of obstruction of the small intestine, many of these patients are subjected to multiple operations before the correct diagnosis is established. At operation, the usual finding is only that of dilated intestine without a definable point of obstruction. Recognition of the true state of affairs is vital lest long segments of intestine, perhaps harboring uninflamed jejunal diverticula, be resected, leaving the patient as an intestinal cripple, a circumstance encountered once by the writer. The diagnosis can be established in many instances without operation so long as it is weighed by the physician. *Intestinal pseudo-obstruction should be considered in any patient who has repeated episodes of partial intestinal obstruction, especially in the absence of hernias or with a history of previous operations.* In such a patient, a weak lower esophageal sphincter and disordered motility of the jejunum and/or colon by manometric examination will provide the correct diagnosis, especially if a family history is present.

Operation should generally be avoided in these patients except under very special circumstances in which manometric studies have shown that benefit can be achieved by bypass of specific areas of severe involvement.

ACUTE OBSTRUCTION OF THE LARGE INTESTINE

Obstruction of the large bowel seldom gives rise to such acute symptoms as those met with in acute obstruction of the small bowel. In the normal person the contents of the colon move on much more slowly and are retained much longer than in the small intestine. The acute symptoms in the large bowel are often preceded by milder attacks of subacute obstruction. The main causes of large-bowel obstruction are *cancer of the colon* (70 percent), *diverticulitis* (5 percent), and *volvulus* (10 percent). Fecal impaction in the rectum may also sometimes cause some obstruction with abdominal distention.

Cancer of the colon

If strangulated hernia is excluded, cancer of the large bowel is the commonest cause of intestinal obstruction in persons over middle age. The symptoms are often insidious, and though in most cases an acute attack of obstruction may be the direct cause for the calling in of the surgeon, yet there are many earlier warning signs and symptoms that should put the observer on guard and cause a thorough examination to be made.

SYMPTOMS OF CANCER OF THE COLON
PRIOR TO ACUTE OBSTRUCTION

Diarrhea and the passage of blood and mucus may result from ulceration of the bowel. The occurrence of diarrhea may lead patients to assert that the bowels are regular, whereas the looseness is but secondary to the irritation caused by the constipated feces. Due to the presence of mucus, the condition may wrongly be diagnosed as mucous colitis.

The presence of a tumor, which in the early stages may be freely mobile, is uncommon, since the majority of the cases are of the contracting scirrhoid type with no palpable tumor (Fig. 27).

The growth may adhere to the bladder or the pelvis of the kidney and cause symptoms referable to the urinary organs, or it may cause gastric symptoms owing to adhesion to the stomach.

Pericolitis, or inflammation of the tissues around the colon, may be the first symptom of note. A local abscess may form and mask the primary condition. Sometimes perforation of the bowel into the general peritoneal cavity suddenly takes place, and diffuse and generally fatal peritonitis ensues.

SYMPTOMS OF CANCER OF THE COLON
SUBACUTE OBSTRUCTION

These are the same symptoms as those caused by acute obstruction, but they are all of slighter degree. *Gradually increasing constipation* is often the first abnormality, and if this supervenes in a person over middle age who has previously been perfectly regular as to the bowels, suspicion should be aroused and a thorough investigation carried out. Diarrhea sometimes alternates with the constipation. Occasional attacks of distention and flatulence are common. Peristalsis may sometimes be seen through the abdominal wall, and

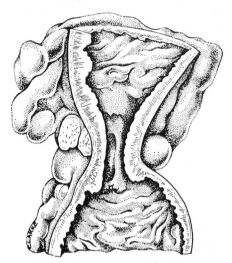

Fig. 27.   The common ring type of scirrhus cancer of the colon that causes intestinal obstruction. (From a specimen in St. Mary's Hospital Museum)

local swelling may subside with a gurgling sound due to passage of flatus through the stricture. Pain is cramplike and due to the peristalsis. Sometimes the patient will describe the pain as traveling across the abdomen and increasing in intensity up to the site where the gurgling occurs and where the obstruction is situated. The pain is often mistaken for indigestion, and the accompanying nausea or sickness is attributed to a bilious attack. This is especially true in obstructions of the right side of the colon.

SYMPTOMS OF ACUTE OBSTRUCTION

Acute obstruction is sometimes, though seldom, the first significant symptom that compels attention. The acute symptoms are, as a rule, precipitated by a hard bolus of food stopping up the lumen, which has been gradually narrowing for a long time. Plumstones or similar hard objects may suffice to complete the obstruction. The contents of the left side of the colon are more firm, and the obstruction is more common in that part. The main symptoms are attacks of *colicky pain*, gradually *increasing distention*, and the *failure to pass flatus or feces.*

Since the contents of the cecum and the right half of the colon are more fluid, it is not so common for acute symptoms of obstruction to supervene from a growth in this part. If, however, the ascending colon does get obstructed by a scirrhoid growth, the final onset of acute obstruction may lead to a sudden distention of the cecum that, as long as the ileocecal valve remains intact, will form a painful, rounded, resonant, and probably tender swelling in the right iliac fossa and the hypogastrium.

When the obstruction is in the left side of the colon and the ileocecal valve functions well, it will be possible to locate the obstruction by noting where the distention suddenly ceases. This can be more clearly shown by a plain radiograph (Plate 13).

When, however, *the ileocecal valve loses its efficiency,* the lower ileum soon becomes distended, and it may be very difficult to determine the exact point of obstruction, for the symptoms of small-gut obstruction are added to those of obstruction of the large bowel. Severe pain, frequent vomiting, and increasing distention present great difficulties, and even an X-ray may not unravel the problem. *Whenever such a differentiation is considered, an emergency barium enema is of the greatest diagnostic value and should always be done.*

Distention, vomiting, pain, and constipation occurring in an elderly person, without any evidence of peritonitis, are generally due to cancer of the large bowel, volvulus, diverticulitis, or, rarely, intussusception or uremia. Inasmuch as cancer of the rectum may at a late stage lead to complete and acute obstruction, it is always necessary to make a careful rectal examination. The urine must be examined to exclude renal disease.

Diverticulitis of the colon

Diverticulitis, when it causes obstruction, produces symptoms indistinguishable from those of cancer of the bowel. Chronic inflammation may cause subacute and eventually acute obstruction of the colon. The symptoms will at first be troublesome constipation, perhaps alternating with attacks of diarrhea, local pain and flatulence, and sometimes hemorrhage. A barium enema may reveal a condition of diverticulosis.

Minor degrees of obstruction may be relieved by medical means, but when acute obstruction supervenes with great abdominal distention, the need for a surgeon will become imperative. Unless it is known that the patient has been for some time suffering from diverticulitis, it may be impossible to make certain whether the obstruction of the colon is due to cancer or diverticulitis before the abdomen is opened; even after the abdomen has been opened and the local lesion examined, there may be great difficulty in determining whether it is cancer, diverticulitis, or a combination of the two conditions.

Volvulus

A volvulus is caused by the twisting of a coil of gut on its own axis to such an extent that the lumen of the bowel is occluded. If the twist is severe enough, the veins and then the arteries may also be compressed, resulting in gangrene of the affected loop.

Volvulus of the large intestine occurs usually in one of two places—the sigmoid and cecal regions. The sigmoid is by far the more common location, owing to the fact that the sigmoid mesocolon is long and the base of attachment narrow, so that twisting of the loop more readily occurs. Ileocecal volvulus is more rare.

VOLVULUS OF THE SIGMOID COLON

Sigmoid volvulus causes symptoms of acute or subacute intestinal obstruction of the large-bowel type. There is usually a preliminary period during which attacks of abdominal pain and constipation may occur.

The *acute attack* is signaled by absolute constipation, acute abdominal pain, and *rapid distention of the abdomen*. If the vessels of the loop are completely occluded, gangrene of the loop quickly occurs and peritonitis ensues rapidly, but gangrene of the colon is usually late because occlusion of the vessels tends to occur slowly. Fortunately, gangrene of the sigmoid has become less common in recent years.

Subacute volvulus is distinguished by abdominal pain (chiefly referred to the umbilical and hypogastric zones), constipation, and gradually increasing distention. The distended sigmoid coil may stand out sometimes in the lower abdomen like the segment of a large pneumatic tire, but later the whole abdomen will become generally distended.

The condition should be suspected when early and great distention supervenes in a case of acute obstruction. Localization of the distention to the hypogastrium at the outset, or the standing out of one large coil, may point clearly to a sigmoid volvulus. Though the pain of sigmoid volvulus is usually hypogastric and not severe, I have known a twisted sigmoid in a young woman to cause such severe pain in the lower lumbar region posteriorly that it was thought that the pain might be due to acute dysmenorrhea. Considerable help can be gained from examination of a plain X-ray of the abdomen when the one distended coil may be plainly visible (Plate 10). Barium enema should be done in all suspected cases. Not only is it diagnostic by demonstrating the so-called bird beak at the apex of the twist, but it is often therapeutic in untwisting a non-gangrenous loop (Plate 11). An alternative therapeutic maneuver is a sigmoidoscopy.

ILEOCECAL OR CECAL VOLVULUS

This gives rise to symptoms similar to, but more acute than, those described under obstruction of the lower end of the small gut, but in addition there will be localized distention due to the dilated cecum, observable either in the epigastrium or the right side of the abdomen. Pain is very severe and vomiting is a prominent symptom. Later, the general distention masks the local cecal dilatation.

X-rays may show a characteristic picture, with the twisted loop oriented toward the left hypochondrium (Plate 9).

VOLVULUS OF THE TRANSVERSE COLON

This is seldom seen except after major abdominal operations in which the intestines had to be displaced and the colon may have been twisted in replacement. I have known it to occur after abdominoperineal excision of the rectum and after hysterectomy. Some days after the operation the abdomen becomes greatly distended and the bowels are not opened, yet there is no sign of peritonitis or of paralysis of the intestines. Reopening of the abdomen is necessary to correct the volvulus.

Surgeons in certain parts of the tropics need to remember that volvulus, of both the large and the small bowel, is relatively much more common than in temperate climes.

Differential diagnosis of obstruction of the large gut

In every case it is necessary first to examine all the hernial apertures. There are four conditions that may deceive the observer and cause him erroneously to believe that he is dealing with a case of primary obstruction of the large bowel. They are:

1. Colitis with distention.
2. Uremia.
3. Peritonitis.
4. Reflex paralysis of the colon.

COLITIS WITH ATONIC DISTENTION OF THE INFLAMED BOWEL

There are cases of severe ulcerative colitis and Crohn's colitis in which, either as a direct result of the ulceration or as a consequence of the toxemia, the large bowel becomes enormously distended and atonic, so that organic obstruction is diagnosed. Toxic dilatation of the colon has also been reported in patients with amebic dysentery, *Clostridium difficile* colitis, and salmonellosis. If seen at a late stage for the first time, it may be impossible to distinguish between toxic megacolon and obstruction, but the long history of symptoms pointing to ulceration of the colon (diarrhea, with passage of blood and mucus) in the one case, and the usual preceding history of subacute obstruction in the other, may help determine the condition. In toxic dilatation of the colon, the distention is usually most marked in the transverse colon. In ulcerative colitis the obstruction is never complete. Rotation of the position of the patient at frequent intervals may allow some passage of flatus and decompression of the colon.

UREMIA

Severe abdominal distention and vomiting are sometimes seen in uremia, and unless a practice is made of examining the urine and blood in every case of supposed intestinal obstruction in middle-aged or older persons, serious mistakes will be made. The estimation of the blood pressure and percussion of the cardiac dullness may throw light on doubtful cases. Symptoms indistinguishable from those of intestinal obstruction may occur.

PERITONITIS

There are some forms of peritonitis that are accompanied by very slight rigidity of the abdominal wall, and there are some abdominal walls in obese, flabby subjects that are almost incapable of becoming rigid on account of the weak, fat-infiltrated muscles. In such patients the distention and vomiting of peritonitis may be mistaken for mechanical obstruction. The late stages of peritonitis are accompanied by a paralytic obstruction of the bowels, and the later stages of intestinal obstruction are frequently accompanied by peritonitis due to organisms escaping through patches of local gangrene or ill-nourished areas of the gut.

In the early stage of both conditions the diagnosis is usually clear on considering the history and symptoms, but in the later stages it may be impossible to differentiate them.

REFLEX PARALYSIS OF THE COLON (COLONIC ILEUS)

There is a deceptive form of paralytic distention of the colon that on several occasions I have known to simulate obstruction of the large bowel. It appears to be a reflex result of an acute inflammation somewhere in the abdomen, and in three cases that I can recall, it was a secondary effect of acute cholecystitis and masked the primary condition. It may develop spontaneously or, more often, after some retroperitoneal operations. The ascending transverse and descending colons may be distended, and it may be difficult to get the bowels to act. These symptoms, with the pain and vomiting, are sufficient to divert attention from the true cause unless special care is taken.

A *barium enema will always distinguish this condition and the other nonobstructive colonic dilatations from true mechanical occlusion.* A colonoscopy can also be both diagnostic and therapeutic. This examination should be omitted in cases of colitis with distention, where perforation is easily possible.

Paralytic obstruction of the intestines (adynamic ileus)

So far we have been dealing with mechanical obstruction of the gut. There is another form of obstruction of the bowel that gives rise to distention and vomiting in which no mechanical obstruction is present and the condition is due to a paralysis of the musculature of the bowel. This condition may occur as a terminal phase in general peritonitis or may ensue after severe chest injuries, after myocardial infarction or during pneumonia, or after operations on the spine or abdomen. The abdomen becomes distended *but on auscultation no borborygmi will be heard.* The whole abdomen tends to be silent but regurgitation of gastric contents takes place, more in the nature of an overflow than due to active movements of the intestine. Both the small and large intestines are involved in the process, so that on an X-ray, gaseous distention involving both will be seen. Apart from keeping the stomach empty (by tube and suction) and maintaining the body fluids and electrolytes, this condition does not call for operation. When, however, the patient is first seen by the surgeon in this condition, it is not always easy to be sure whether the final paralytic state of the bowel may not have been a terminal state following mechanical obstruction. In such cases, a careful history may permit a correct judgment to be made.

14. Intussusception and other causes of obstruction

Intussusception

Intussusception or invagination of the intestine is the most common abdominal emergency in children under two years. It constitutes 15 percent of all cases of intestinal obstruction. In most parts of the world it is much less common in later childhood and adult life, only 30 percent of cases occurring after the second year of life. (This does not hold good for some parts of Africa and India.) The catastrophe is all the more unexpected in that it usually attacks the most healthy-looking and well-nourished babies. The condition consists in the invagination of one portion of intestine into the portion next to it. Commonly, if not invariably, the invaginated part (intussusceptum) enters the part below (intussuscipiens). Clearly, the most anatomically favorable part for such an occurrence is in the ileocecal region, where the narrow ileum can readily enter the lax cecum, and in actual clinical experience this is the most common place for the condition to start (Fig. 28). There are three varieties of intussusception—enteric, enterocolic, and colic. Enteric, where the small intestine alone is involved, is uncommon; colic, in which the colon alone is affected, is less rare but not very common; enterocolic is the most common variety. The enterocolic type is subdivided into ileocecal, in which the apex of the invaginated part is

Ileocecal

Colic

Enteric

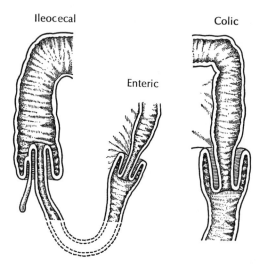

Fig. 28. Types of intussusception: (1) ileocecal, (2) enteric (which is termed "ileocolic" if it progresses beyond the ileocecal valve), (3) colic.

the ileocecal valve, and ileocolic, in which a part of the gut near the end of the ileum forms the advancing apex.

In certain parts of the tropics, for example, central Africa and India, intussusception is more common in adults. It is usually of the ileocecal type, does not lead to complete obstruction, and may not cause any bleeding per rectum. Colicky pain and the presence of a tumor are the two main symptoms.

In the case of the most common form—the enterocolic—as the end of the ileum is invaginated into the colon a portion of the mesentery goes with it, and constriction, and later strangulation, of the vessels occur, causing edema of the gut wall with intestinal hemorrhage and finally gangrene. The irritation caused by the intruded gut leads to excessive secretion of mucus. The part of the gut that first becomes invaginated remains at the apex of the advancing portion, and it progresses at the expense of the en-sheathing layer (intussuscipiens). The apex is sometimes extruded at the anus. If left untreated, intussusception ends in one of two ways. Most commonly the intestinal lumen is gradually occluded, and acute intestinal obstruction and death from toxemia or peritonitis result; more rarely, the invaginated part becomes completely gangrenous and passes per anum as a large slough. In preoperative days many such cases were recorded, but since it was a comparative rarity for infants to recover from the condition, such a fortunate event as a natural cure by gangrene of the intussuscepted part was probably more often recorded.

The cause of intussusception in infants is unknown. It commonly occurs in infants at the weaning time, when there are likely to be occasional portions of undigestible solid food taken by the well-nourished baby. In later life, tumors of the gut wall or Meckel's diverticulum are commonly the cause of the condition.

SYMPTOMS AND SIGNS

The symptoms and signs of intussusception are usually characteristic. They comprise a few or many of the following according to the stage at which the case is seen:

1. Abdominal pain.
2. Passage of blood and mucus per anum.
3. Vomiting.
4. An abdominal swelling.
5. Shock.
6. Visible peristalsis.
7. Absence of the cecum from the right iliac fossa.
8. Constipation.
9. Tenesmus.
10. Distention of the abdomen.
11. Fever (sometimes).
12. Obstruction and peritonitis (late stage).

Abdominal pain. The onset is usually with a *fit of screaming*—the infant's method of indicating pain. The legs are drawn up during the screaming attacks. The pain is very severe but is not continuous and corresponds to the violent peristaltic contraction of the gut. Between the bouts of pain, the child may lie quiet but often has an apprehensive look. Less often, the child does not scream or show any sign of abdominal pain other than pallor and drawing up of the legs or restlessly rolling over on the bed.

Blood and mucus. At a period varying according to the site of the invagination—later if it starts in the ileum, earlier if in the transverse colon—*blood and mucus are passed per anum.* This usually occurs within a few hours. The blood is often quite slight in amount, and it is seldom copious. Slimy mucus is mixed with the blood, and not infrequently, some brown or yellow fecal material may also be passed. Cases do occur, however, in which no blood is passed per anum before the child comes under observation, sometimes up to as long as twenty-four or forty-eight hours after onset of the pain. The practitioner must be prepared to diagnose intussusception before blood has been passed per anum.

Vomiting. This generally occurs, but it is not severe at first. It is never a serious feature until the late stages when complete obstruction has ensued or peritonitis has developed. The contents of the stomach are returned, any liquid taken is not retained and later there may be bilious vomit, but feculent vomit is rare.

Abdominal tumor. By the time blood appears at the anus, *a tumor* is present in the abdomen. It is caused by the invaginated gut and is felt either in the right loin, right hypochondrium, epigastrium, left hypochondrium, left lumbar region, or left iliac region, according to the advance made by the intussusception through the colon. The tumor is oval in shape and has often been compared—quite aptly— to a sausage. Sometimes the swelling becomes harder, the change corresponding to the muscular peristaltic contraction. Frequently it is easy to feel the swelling, but in many cases relaxation of the abdominal wall is insufficient to allow recognition of the mass. *The hiding of the intussusception by the overhanging liver is responsible for many failures to detect the tumor.*

Shock. The severity of the pain is shown by the extreme facial pallor, the dilated pupils, and the anxious appearance of the child. True circulatory shock is late, and the mortality is much higher when it occurs. The diagnosis should be established before it occurs.

Visible peristalsis. Peristalsis may sometimes be seen through the abdominal wall, and the simultaneous hardening of the tumor has been referred to above.

Empty right iliac fossa. In the common enterocolic variety, almost from the beginning of the illness *the right iliac fossa will appear empty on palpation* due to the taking up of the cecum into the advancing invagination (signe de Dance) (Fig. 29).

Bowels. Constipation is by no means always absolute. Exceptionally, an intussusception may be present and yet the bowels may open fairly normally; not uncommonly, fecal material may be mingled with the blood and mucus that come away. As the condition advances, however, the obstruction increases and ultimately becomes absolute. One must therefore be prepared sometimes to diagnose intussusception in the absence of absolute constipation.

Tenesmus. As the intussusception approaches the rectum, *tenesmus* may be indicated by the constant straining efforts of the infant. At this stage the congested apex may sometimes be felt on rectal examination. In some cases the *apex* of the intussusception *may protrude through the anus* as a red, congested, fleshy mass.

Distention. In late or neglected cases the increasing obstruction of the lumen of the gut results in *abdominal distention* and increased frequency of vomiting.

Obstruction and peritonitis. The final stage is that of complete intestinal obstruction and peritonitis due to gangrene of the de-

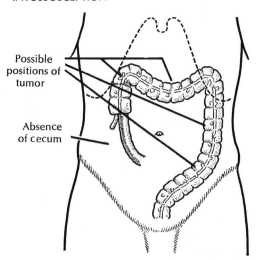

Possible
positions of
tumor

Absence
of cecum

Fig. 29. The possible positions of abdominal tumor and how the cecum is absent
from the right iliac fossa in intussusception. (The latter is a valuable diagnostic point
in distinguishing intussusception from colitis.)

vitalized gut and infection of the peritoneum. Repeated vomiting,
signs of toxemia, and exhaustion end the scene.

In addition to the above symptoms, fever of up to 100°F or even
higher may be present as early as the first twenty-four hours.

DIAGNOSIS

It is usually not difficult to diagnose an intussusception. The age of
the child, the previous good health and sudden onset of acute pain
coming on in bouts, which cause severe temporary shock, the
passage of blood and mucus per anum, and the presence of a
sausage-shaped swelling in the abdomen are sufficiently characteristic to admit of no doubt. The cases of real doubt are those in
which, when the doctor sees the patient, the attacks of pain may
be quiescent and no tumor can be felt. In such cases, if the history is
at all suggestive or characteristic, and blood and mucus have
been passed, a barium enema should be done. By this means
the diagnosis can be definitely made, and at the same time the

enema will help to reduce any intussusception that may be present (Plate 8).

DIFFERENTIAL DIAGNOSIS

In the early stages the condition must be distinguished from:

1. Simple colic.
2. Colitis and enterocolitis.
3. Rectal polypus.

In the later stages one must exclude:

1. Prolapsed anus and rectum.
2. Other causes of obstruction and peritonitis.
3. Henoch's purpura.

Simple colic. With simple colic the evidence of pain is not so outstanding, nor is shock so extreme. No lump can be felt in the abdomen, and no blood is passed per anum. Pain usually ceases when the bowels are emptied.

Colitis and enterocolitis. These furnish the main difficulty in diagnosis in young infants, among whom acute enterocolitis is not uncommon. Colitis is sometimes accompanied by the passage of blood and mucus per rectum. The chief distinguishing features are as follows:

1. In colitis there is usually a stage of preliminary diarrhea unaccompanied by blood.
2. The infants who readily fall victims to colitis are frequently illnourished. Intussusception usually attacks well-nourished obese infants.
3. In colitis there are more frequent stools, as a rule containing more fecal material than in cases of intussusception.

4. In colitis there is no abdominal tumor to be felt.
5. The cecum can be felt in the right iliac fossa, and possibly gurgling may be elicited by pressing on it, or borborygmi may be heard on auscultation. There is not that emptiness that is so noticeable on palpating the fossa in most cases of intussusception.
6. The crises of pain are not usually so severe in colitis.
7. In colitis there is sometimes tenderness along the whole course of the colon.

Obstructive symptoms and distention are not so common in colitis. Tenesmus may be present in both cases. There may be an epidemic of similar cases that may help in the diagnosis of colitis. As mentioned above, a radiograph after a barium enema will always settle the diagnosis. Two or three degrees of fever may be present in each condition, so this symptom cannot be used as a differentiating feature.

Prolapsed anus and rectum. Cases in which the apex of the invagination protrudes through the anus have to be distinguished from prolapse of the rectum. In the latter a ring of prolapsed mucous membrane is seen around a central opening, and the finger can be inserted for only a short distance between the mucosa and the external sphincter; in a prolapsed intussusception the opening of the protruding portion is toward the posterior aspect of the projection, and the finger can easily be inserted between the anterior or lateral portions of the projection and the anal sphincter. Any intussusception that has advanced to the anus will usually be accompanied by considerable distention and symptoms of intestinal obstruction.

Obstruction and peritonitis. The late stages of an intussusception in which the apex has not advanced as far as the rectum, and that is accompanied by advanced symptoms of intestinal obstruction or peritonitis (i.e., distention, frequent vomiting, toxemia, and collapse), can only be diagnosed from the other causes of those symptoms by the history of onset and the previous course of the disease.

Henoch's purpura. In children who have passed infancy, and occasionally in infants, intussusception has to be distinguished from *Henoch's purpura,* a disease characterized by abdominal pain, vomiting, and the passage of blood per anum and frequently accompanied by arthritis and an eruption of purpuric spots. The bleeding from the gut is due to an effusion of blood into the walls of the intestine. The youngest child in the series described by Henoch was four years old, so the age incidence of the two diseases may be a help in diagnosis. In doubtful cases a very thorough search must be made for purpuric spots or joint affections. Very rarely have the two conditions been coexistent. Here again, a barium enema may clinch the diagnosis.

SUBACUTE AND CHRONIC INTUSSUSCEPTION

There are some cases of chronic intussusception that are accompanied by slight signs of intestinal obstruction but progress steadily with repeated attacks of pain, sometimes at considerable intervals, until a final serious attack of obstruction occurs. In these cases there may be normal or almost normal fecal motions until the final attack, and the observer is very likely to be misled by the chronicity or intermittence of the symptoms. I have known such a case taken for *tuberculous peritonitis* and *enteritis.* There were loose fecal motions and an epigastric swelling thought to be rolled-up omentum, but in reality an intussusception had occurred. In these subacute cases the help afforded by radiography is of the utmost value.

INTUSSUSCEPTION OF THE PELVIC COLON IN OLD PEOPLE

Intussusception is very rare in old people, but when it does occur, it generally affects the sigmoidorectal region. This leads to frequent hypogastric pains and tenesmus, while mucus, and later blood, are passed through a rather patulous anus. Rectal examination easily demonstrates the edematous apex of the intussusception, which is seldom more than a few inches long. A malignant growth or polyp may sometimes form the apex. In parts of Africa and in India, intussusception is more common in adults than in infants.

Intestinal obstruction in newborn infants

Intestinal obstruction in infants and young children is always serious and usually dangerous.

Newborn babes may have complete obliteration of the lumen of some part of the small or large intestine, or they may present obstruction due to malrotation of the intestinal loop. The obstruction may be in the duodenum, jejunum, ileum, or large bowel. Sometimes there is not a complete obstruction but a narrowing or stenosis; in these cases the symptoms will be more gradual in onset.

If the obstruction is in the duodenum or upper ileum, the symptoms begin within forty-eight hours of birth with the onset of vomiting, which continues and becomes persistent. The vomit invariably contains bile. All the liquid food taken by mouth is quickly returned, and the stomach and upper duodenum become distended and cause fullness in the epigastrium. The lower abdomen is normal. There may be visible gastric peristalsis. The babe is apathetic, does not show any desire to take anything by mouth, and rapidly wastes away and becomes dehydrated. A straight X-ray will show much gas in the stomach and duodenum, but not beyond. The clinical picture and the characteristic X-ray picture usually make the diagnosis fairly clear. *When the so-called double bubble of the stomach and first part of the duodenum is seen, the obstructing mechanism is duodenal atresia, annular pancreas, or duodenal bands associated with malrotations.*

When the obstruction is lower down in the intestine, in the ileum, the symptoms come on more slowly but quite definitely.

The coils of intestine will gradually distend and may form a ladder pattern. The vomiting, which comes on several days after birth, will change from bilious to brownish and will have a feculent odor. An X-ray will show coils of distended small bowel but will show no gas in the colon. The obstructing mechanism is usually an atresia, although volvulus of the midgut and meconium ileus should be considered.

An imperforate anus is usually noted at birth, but if there is an occlusion of the colon higher up, several days may pass before serious symptoms occur. The gradually increasing abdominal distention, and perhaps the onset of vomiting, will cause alarm, and the absence of normal stools will certainly cause further investigation to be made.

In atresia of the small or large gut, stools may be passed that, though not normal, may not at once call attention to the serious abnormality. Fortunately, there is a simple test that will determine whether or not there is a complete obstruction. The intestine of all newborn babes contains some vernix caseosa that has been swallowed. This should appear in the stools if no obstruction is present. The presence of the cornified cells of the vernix caseosa can be identified readily by Farber's test. A smear of the contents of the bowel is made on a glass slide and is stained with gentian violet; the slide is then washed with running water and decolorized with acid alcohol. The cornified cells of the vernix retain the stain and can easily be recognized.

MECONIUM ILEUS

Meconium ileus is a form of intestinal obstruction that, until comparatively recently, was almost invariably fatal; during the last twenty years, however, a method of treatment has been evolved that saves the lives of a number of these patients.

Meconium ileus is a condition, usually associated with fibrocystic disease of the pancreas, in which the meconium in the lower part of the intestine becomes inspissated to such an extent that the lumen of the bowel is completely occluded. The meconium is thick and sticky and cannot easily be shifted along the bowel. Commonly it is found plugging the last 10, 20, or even 30 cm of ileum, which is considerably dilated and greatly thickened. Vomiting starts a day or two after birth, and there is moderate distention. The stools may not give much definite information, but a straight X-ray will show the coils of intestine varying greatly in size, usually without air-fluid levels. Inspissated meconium sometimes has a granular appearance (on a radiograph) that is never seen in other forms of obstruction. In uncomplicated cases, a Gastrografin enema will usually make the diagnosis and clear the obstruction without resort to open operation. In any neonatal patient with obstruction of the small intestine, a contrast enema usually demonstrates a so-called microcolon. This should not be misinterpreted as an anatomical narrowing. Rather, it represents a functional collapse of the colon that is easily reversed either by use alone or by gently distending enemas if deemed necessary. Treatment of the pancreatic deficiency has to be continued afterward.

HYPERTROPHIC PYLORIC STENOSIS

This is an emergency that usually arises in the second month of life. The symptoms begin about the sixth week after birth with vomiting that gradually becomes persistent. The vomit does not, as a rule, contain any bile. Gastric peristalsis may be seen, and usually a small tumor, the so-called olive or olive pit, can be felt in the upper abdomen, lying under the liver in the right hypochondrium or epigastrium. The symptoms are usually sufficiently characteristic to make diagnosis straightforward. An ultrasonic examination may be needed to confirm the presence of the olive.

INTESTINAL OBSTRUCTION DUE TO MALROTATION OF THE MIDGUT

At an early period of normal embryonic development, as the midgut returns to the abdominal cavity from the umbilicus, there occurs a counterclockwise rotation of the midgut so that the cecum and that part of the colon that forms the hepatic flexure take their normal position on the right-hand side of the abdomen.

Occasionally this rotation does not occur, and in some infants attacks of intestinal obstruction may follow. It has, however, been shown comparatively recently that this anomaly may be present and cause symptoms in adults.

In neonatal life, malrotation produces intestinal obstruction either because of bands that extend from the incompletely rotated cecum to the posterior abdominal wall, thus binding and occluding the duodenum, or because of volvulus of the midgut. The latter occurs because there is no fixation of the mesentery of the small intestine, thus allowing several rotations about the axis of the superior mesenteric vessels and consequent gangrene.

The abnormal symptoms presented in adult life are those of acute or subacute obstruction of the intestine because they usually arise as a result of volvulus of the midgut, rather than because of the duodenal bands. The attacks may occur from early life up to as late as the sixth decade, and the interval between the recurrent bouts of obstruction varies from a few weeks to several months. As a rule the attacks are self-limited and last from one to three days. The symptoms are fairly definite. Each attack begins with central abdominal pain accompanied by vomiting and ending with diarrhea, indicating that the obstruction has righted itself. Between the attacks there may be no abnormal physical signs or symptoms, but a history of such attacks should lead the observer to consult a radiologist with a view to a special X-ray examination to detect malrotation of the midgut.

15. The early diagnosis of strangulated and obstructed hernias

A strangulated hernia is one of the commonest and most dangerous forms of intestinal obstruction. The hernial orifices through which the abdominal contents protrude have for the most part hard, fibrous edges that quickly cause either local necrosis of the gut at the site of pressure or interference with the blood supply in the accompanying mesentery, with a consequent gangrene of the gut. It is often very difficult to make certain whether a hernia is merely obstructed or whether it is strangulated, for pain and constipation are usually present in both cases.

The symptoms of a strangulated hernia are those of intestinal obstruction with the addition of a painful, tender, and often tense swelling in one of the hernial regions. In certain cases, however, there may be little or no local tenderness to call attention to the hernia. The obstructive symptoms will vary according to the nature of the parts strangulated in the sac. If jejunum is caught in the hernia, very acute symptoms ensue; if ileum is obstructed, less acute manifestations result; while with large bowel alone in the sac, the symptoms are usually subacute but nonetheless serious. When omentum alone is strangulated or if a Richter's hernia is present, there will be pain, constipation, nausea, and sometimes vomiting, but the obstruction of the gut is never complete.

Shock is a variable factor but may be severe in late strangulation.

The vomiting gradually becomes feculent when the small intestine is completely obstructed in its lower portions.

Since, apart from operation, it is nearly impossible to make certain that there is no gut in the sac, and inasmuch as mechanical obstruction to the bowel is ultimately as dangerous as strangulation, *it is well to treat all painful, tense hernias as if they were strangulated.* We consider that, except in the case of very easily reducible swellings, all painful hernias, accompanied by abdominal pain and other symptoms of obstruction, should be treated as strangulated hernias by operation without attempting taxis. If an anesthetic has to be given for reduction, it is much better to reduce by open operation. Fomentations and icebags should be regarded as causing unwise and dangerous delay.

There are a few practical diagnostic points that may be considered with each of the several varieties of hernia.

Strangulated inguinal hernia

It is usually a very easy matter to diagnose a strangulated inguinal hernia. The patient has usually been aware of the existence of the hernia for some time and may have been wearing a truss. The sudden coming down of the rupture, accompanied by abdominal pain, vomiting, and tenseness and tenderness in the local swelling, makes a clear picture. But mistakes are possible in the following conditions:

Vomiting due to some other condition. This may in the presence of an unreduced inguinal hernia give rise to a suspicion of strangulation. The vomiting of pregnancy or that at the onset of appendicitis may lead to this mistake. But in such cases *the sac is usually neither tender nor tense, and the contents may be easily reducible.* The problem is made more complex when a patient with a chronically incarcerated and irreducible hernia suffers an abdominal catastrophe of other cause, but in this case the sac will not be tense. When a catastrophe, such as the perforation of a peptic ulcer or rupture of the gallbladder, happens to occur in a patient who has an unreduced hernia, there will be tenderness over the hernial site, and the diagnosis has to be made by considering the history and other abdominal signs and symptoms.

Torsion or inflammation of an undescended inguinal testis. Occasionally this may closely simulate a strangulated hernia. The absence of the testicle from the scrotum on the affected side should make one consider the possibility of this condition. In torsion of the testicle the local pain is very severe, and vomiting will begin early but will never become feculent. Though constipation may be a symptom, enemas will produce satisfactory results.

When a metastasis of mumps occurs in an inguinal testis—a very rare occurrence—it might lead to a diagnosis of strangulated hernia, but there would be no intestinal obstruction.

It is to be remembered that a retained (undescended) testis is often associated with a hernia of the interstitial variety.

Inflamed inguinal or iliac glands. With these the swelling is usually more diffuse and fixed, and there may be redness and edema of the skin and subcutaneous tissues. Vomiting and intra-abdominal pain will be absent or slight, and fever is sure to be present. Occasionally, a distinction cannot be made by means other than operation.

Acute hydrocele of the cord. Although rare, this has been known to simulate a strangulated hernia, but the painful, tense inguinal swelling of an acute hydrocele does not cause so severe vomiting, and never feculent vomiting, nor is there any evidence of intestinal obstruction.

Thrombosis of veins of the spermatic cord. Thrombosis or suppurative phlebitis of the veins of the spermatic cord causes a painful swelling of the cord and its surroundings, but with these conditions the testicle is swollen and tender and will give the clue to the diagnosis.

Subperitoneal fibroid. On several occasions I have known a subperitoneal fibroid in pregnant women to cause a swelling in the inguinal region, and lead to suspicion of strangulated hernia on account of the coincident vomiting of pregnancy. Careful examination enables one to feel the tumor moving separately from the abdominal wall, and, of course, the vomiting never becomes feculent and the tumor is not usually tender.

Strangulated femoral hernia

A strangulated femoral hernia gives rise to more mistakes in diagnosis than a strangulated inguinal hernia. The hernia is often very small and may easily escape notice in the thick fat often present in the saphenous region. Sometimes only a small knuckle of gut comprising a small portion of the circumference of the bowel may be caught in the femoral canal, and scarcely any projection may be felt in the thigh. When the hernia is large, it consists of a rounded fundus and a narrow neck or stalk that permits free and often *painless* side-to-side movement of the fundal part. This absence of fixity of the sac may lead the observer to think that there is no strangulation.

A strangulated femoral hernia might be simulated by:

1. An inflamed inguinal lymph gland.
2. Thrombosis of a saphenous varix.
3. An inflamed appendix in a femoral hernial sac.
4. A tense, painful hydrocele of a strangulated inguinal hernia.

A strangulated inguinal hernia has often been thought to be present when, in fact, the trouble is in a femoral sac.

Inflamed inguinal lymph gland. An inflamed gland is usually more fixed, gives rise to more edema and possible redness of the overlying parts, and usually results from a primary cause that may be detected on the corresponding thigh or anoperineal region. This is not always the case, however, and the distinction from a hernia may be impossible. If this proves to be the case, an operation should be done. Vomiting and intestinal obstruction are, of course, absent.

Thrombosed saphenous varix. A thrombosed saphenous varix does not ordinarily give rise to vomiting or abdominal pain. If the thrombosis extends up to the iliac vein, there will be pain and tenderness in the iliac region.

Inflamed appendix. An inflamed appendix in a femoral hernial sac cannot be distinguished definitely from a strangulated omental femoral hernia before operation, though by the history of previous attacks of appendicitis it might be suspected. If feculent vomiting occurs, it would, of course, be clear that bowel was strangled in the sac.

Strangulated inguinal hernia. This is distinguished from a femoral hernia by noting that the swelling comes out of the abdomen medially to the pubic spine and above Poupart's ligament.

In general, any swelling in the region of the femoral canal that suddenly appears or becomes larger and more painful, and is accompanied by *vomiting* or *nausea,* should be considered as a femoral hernia in need of immediate operation. A small femoral hernia containing strangulated gut may sometimes give rise to very slight local pain or tenderness. In such cases the general abdominal pain and vomiting may dominate the scene, and the small hernia may be overlooked.

In those rare cases in which a pouch of the cecum becomes tightly gripped in a femoral hernia, there may be no abdominal pain, no vomiting, and no intestinal obstruction, even though the herniated portion becomes gangrenous. This fact makes it all the more necessary to explore any doubtful tender lump in the femoral region. Whether one is dealing with a strangulated inguinal or femoral hernia, the distinction between the two is of intellectual but not practical interest. Both require urgent operation.

Obstructed and strangulated umbilical, ventral, and obturator hernias

UMBILICAL HERNIA

Umbilical or paraumbilical hernia is more common in women and is chiefly seen in obese persons (Fig. 30).

An umbilical hernia usually contains omentum and frequently large bowel. Small intestine is not so commonly found in the sac. When obstruction or strangulation occurs, the symptoms are therefore much more likely to be subacute and to partake more of the character of large-bowel obstruction. The two common mistakes made in regard to umbilical hernia are (1) to overlook a small hernia lying deeply embedded in fat and (2) to think that the hernia is not strangulated because the symptoms are not very acute. The fact that the percentage mortality for operations on strangulated umbilical hernia is three times that for strangulated inguinal and femoral hernias shows the serious need for more early diagnosis and interference.

It is quite common for a patient to have several attacks of obstruction before the more serious strangulation attack occurs. The fact that on previous occasions the obstruction has been overcome by aperients may lead to an erroneous opinion that the same will occur again. It is sometimes only by the most serious symptoms of strangulation (gangrene of gut or omentum, fecal abscess, even gangrene of the skin overlying the hernia) that the patient and medical attendant realize the extremely serious nature of the case.

Diagnosis. The diagnosis of an obstructed umbilical hernia is made on the occurrence of abdominal pain, vomiting, constipation, and

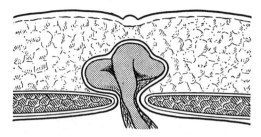

Fig. 30. How an umbilical hernia may be embedded in, and hidden by, a fat abdominal wall.

local tenderness on pressure over the swelling, which can always be felt deep in the fat, even if it does not bulge obviously on the surface in the region of the umbilicus. If the small gut is obstructed in the sac, the symptoms will be correspondingly more acute and vomiting will occur and may become feculent.

It is often difficult to say before opening the sac whether one is dealing with simple obstruction or with strangulation.

VENTRAL HERNIA

The same general remarks apply here as in the case of umbilical hernia, except that small bowel is more commonly found in the sac, and consequently, acute symptoms are more frequently observed. Abdominal pain, vomiting, constipation, and local tenderness over the site of a ventral hernia are sufficient to indicate the need for operation.

OBTURATOR HERNIA

An obturator hernia is one that forces its way along the fibro-osseous canals through which pass the obturator vessels and nerve. The hernia is always small and should almost be classed among the internal hernias, though occasionally, when strangulation occurs, there may be tenderness and a little fullness in the upper adductor region of the corresponding thigh. The condition is most common in wasted, elderly women.

The symptoms of strangulation are those of obstruction of the small intestine, *often of the Richter type.* The only local symptom may be some pain radiating down the inner side of the thigh along the distribution of the obturator nerve.

The thigh-rotation test may be positive, and indeed, any movement of the thigh that brings into action the obturator muscles may cause pain. *Rectal examination is important, for it may reveal a tender, palpable mass* in the region of the obturator canal.

16. Acute abdominal symptoms due to vascular lesions

The past twenty years have seen a rapid and notable advance in the surgery of the large blood vessels, and many conditions that were formerly outside the realm of curative surgery have become amenable to such treatment.

The main conditions that need to be noticed are:

1. Mesenteric arterial thrombosis or embolism.
2. Nonocclusive intestinal ischemia.
3. Mesenteric venous thrombosis.
4. Dissecting aneurysm.
5. Leakage or rupture of an abdominal aneurysm.
6. Internal hemorrhage from other causes.

Mesenteric arterial thrombosis or embolism

Mesenteric thrombosis or embolism most often affects the territory of the superior mesenteric artery, and whether the main trunk of the artery is involved proximal or distal to the middle colic artery branch will obviously influence the course of the disease and the extent of intestinal gangrene. A detailed history indicating abdominal angina during the weeks or months before the acute episode is consistent with thrombotic occlusion of the superior mesenteric artery or its branches. On the other hand, a history of recent endocarditis or, even more importantly, the presence of an arrhythmia such as auricular fibrillation points strongly toward embolization. *Any patient with an arrhythmia such as auricular fibrillation who complains of abdominal pain is highly suspected of having embolization to the superior mesenteric artery until proved otherwise.*

SYMPTOMS AND FINDINGS

The symptoms may be remarkably insidious at the onset, and those of thrombosis generally cannot be distinguished from those of embolism. The patient complains initially of mild central cramping abdominal pain, accentuated by food intake and usually accompanied by mild nausea and vomiting. Frequent bowel movements are common and usually contain either grossly or microscopically detectable blood. The abdomen gradually becomes distended, but prominent tenderness and rigidity are absent until late in the course of the disease. Fever, leukocytosis, hypotension, and tachycardia tend to be very late manifestations (too late!), and attempts to establish the correct diagnosis must antedate the time when the true state of affairs is evident even to the most inexperienced observer. Overt peritonitis occurs so late in the course of the disease that it might well be regarded as a premorbid event. *It is little known or appreciated that the early symptoms are present and are relatively mild in 50 percent of cases for three to four days before medical attention is sought.* This fact attests only too well to the deceptively benign early course of this disease. Arteriography is the only certain method of preoperative diagnosis.

Nonocclusive intestinal ischemia

This condition has been recognized with increasing frequency in recent years. It has become clear that both the small and large intestines may undergo partial- or even full-thickness necrosis *in the absence of overt organic vascular occlusion.* The usual patient is one with congestive heart failure who takes digitalis glycosides and who has recently been given intensive diuretic therapy. It has been shown that the use of digitalis derivatives under these circumstances causes a decrease in mesenteric blood flow out of proportion to decrements in other vascular beds. Prolonged shock or cardiopulmonary bypass may also be forerunners of this serious condition.

The entire clinical picture may be indistinguishable from that of organic occlusion of the mesenteric vessels. Large volumes of fluid are usually required to achieve adequate resuscitation. *Any patient who takes digitalis and diuretics and who complains of abdominal pain must be considered to have nonocclusive ischemia until proved otherwise.*

The condition may involve the entire small intestine and colon, often in a patchy distribution. On occasion, however, one encounters segmental disease, which is probably the result of a preexistent, partially occlusive lesion upon which nonocclusive ischemia has been superimposed.

If this condition or organic occlusive ischemia is suspected, angiography must be done. Plain or barium X-rays will demonstrate the nonspecific findings consistent with an ileus but are usually of no greater help. Occasionally, gas bubbles will be seen in the portal venous system on a plain film. Arteriographic demonstration of an organic occlusive lesion will lead to operation, whereas demonstration of diffuse spasm will indicate the need for vasodilator drugs. Even with early recognition, the mortality rate for nonocclusive ischemia is roughly 50 percent at best.

A relatively benign form of the disease is sometimes observed in the splenic flexure or descending colon, and the sigmoid and rectum may also be the site of isolated disease. Cramping abdominal pain, diarrhea, and the passage of bloody mucus together with slight abdominal tenderness over the involved areas are the typical symptoms. Since the disease involves, to the greatest extent, the tissue that is metabolically most active (i.e., the mucosa), a barium enema will usually suffice to establish the diagnosis by demonstrating large, edematous folds, the so-called thumbprinting. X-rays (barium enema) done about one week later will show complete resolution in most cases, but if the injury is sufficient to involve the submucosa, a stricture may develop.

Mesenteric venous thrombosis

This condition is much less common than the two discussed above. It is to be considered when polycythemia vera or severe liver disease is present, and recently the use of contraceptive pills has been implicated as a pathogenetic factor.

The symptoms and signs are similar to those of arterial ischemia, except that the progression of the disease tends to be more rapid. Thus, shock and signs of blood loss are seen earlier than in arterial ischemia. Arteriography will identify the venous nature of the lesion.

DIAGNOSIS OF ARTERIAL OR VENOUS OCCLUSION AND NONOCCLUSIVE ISCHEMIA

Differentiation has to be made from perforated peptic ulcer, hemorrhagic pancreatitis, and dissecting aneurysm, and, in the later stages, from other forms of intestinal obstruction and peritonitis. A help in diagnosis may be provided by abdominal paracentesis. The obtaining of serosanguineous fluid would indicate that the lesion is likely to be either mesenteric thrombosis, nonocclusive intestinal ischemia, strangulation of a coil of intestine, or possibly hemorrhagic pancreatitis. Arteriography is essential in the diagnosis of most cases, so that appropriate therapy may be undertaken.

Dissecting aneurysm

The term "dissecting aneurysm" is applied to that condition in which there is an erosion of the inner lining of an artery through which the blood escapes into the substance of the wall of the artery and courses along within its substance parallel to the mainstream, splitting the arterial wall in its middle muscular coat. The initial lesion is commonly in the upper thoracic aorta. The aorta may be split along the whole of the thoracic trunk, and sometimes also along almost the whole of the abdominal aorta. Sometimes the blood may force its way back into the mainstream, and there have been a few instances in which this has resulted in stoppage of the tearing process. Commonly, however, the process may extend to the common iliac and lead to a partial or complete occlusion of this vessel.

Though this condition was known many years ago, it did not become a recognizable pathological process and had very seldom been diagnosed during life until 1933. Since that date the condition has more often been correctly diagnosed, for in a considerable number of instances, the symptoms are sufficient for a fairly confident recognition.

SYMPTOMS

The main symptom of dissecting aneurysm is an unbearable pain starting in the thorax, sometimes radiating at first up to the neck or through to the back, but gradually extending down to the abdomen and sometimes even to the hip or thigh. The warning has been given that "In any patient with pain in the chest or back of such severity that narcotics fail to relieve it, the diagnosis of a dissecting aneurysm should be considered." When the lesion reaches the abdominal aorta there will be general abdominal pain, and the abdominal wall may become rigid so that a perforated peptic ulcer, or even hemorrhagic pancreatitis or mesenteric thrombosis, may well be suspected.

DIAGNOSIS

Diagnosis will be helped by consideration of the following points. Significant arterial hypertension of prolonged duration is usually a forerunner. The pulses should be felt in all four limbs and, as far as possible, the blood pressure should be measured in each limb. With a dissecting aneurysm there will almost certainly be serious differences between one upper- and one lower-limb pulse according to the position of the lesion. There may be an aortic diastolic murmur. (This will only be of value if the patient is known not to have had such a murmur prior to the illness.) If the renal artery is involved, there may be some hematuria and perhaps a rise in the blood urea nitrogen. The limitation of the blood supply to the parts supplied by affected arteries may lead to numbness, paresthesia, or even hemiplegia or paraplegia if the arteries supplying the spinal cord are affected. The chief distinguishing features are, however, the change in position of the pain from the thorax to the abdomen, the inequalities of the pulses, and sometimes cessation of the pulse in one femoral artery.

Operative relief may be given these patients if they can be conveyed to a center where surgeons experienced in this type of work are available.

Leakage or rupture of an abdominal aneurysm

Most abdominal aortic aneurysms are fusiform in type (rather than saccular) and are caused by atherosclerosis. They almost invariably involve the aorta *below* the origins of the renal arteries (even though they may feel very high by physical examination), and occasionally the process is found in the iliac arteries (Fig. 31). These lesions are often incorrectly referred to as "dissecting aneurysms." Since the two lesions are pathogenetically distinct and require vastly different therapies, clear separation of the two must always be emphasized.

SYMPTOMS AND SIGNS

There may be no symptoms at all before the vessel ruptures. *Pain in and about an aneurysm portends rupture,* and any patient with a known aneurysm and recent abdominal pain should be regarded as being in imminent danger of rupture. When present, the pain prior to rupture is of a throbbing or aching nature. It is located in the

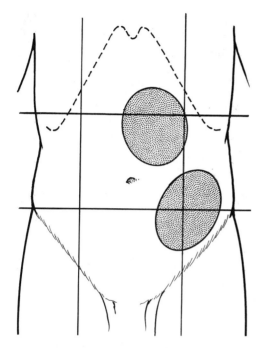

Fig. 31. Common sites for pulsatile masses in patients with leaking or ruptured abdominal aneurysm.

epigastrium or the back (lumbar region). Aneurysms of 6 cm or larger in diameter are more likely to rupture than are smaller lesions.

Several misconceptions about abdominal aortic aneurysms are prevalent:

1. Peripheral pulses are absent, diminished, or abnormal. This very commonly misleads the unknowing physician, who often finds excellent pulses in a patient with an aneurysm. In fact, large popliteal aneurysms (frequently bilateral) often accompany abdominal aortic aneurysms and cause strong, easily palpable pulses.
2. A bruit is often heard over an aneurysm. Again, this is a false but commonly held view.

3. Death is almost immediate following rupture. This is another popular misconception usually held by those who do not care for many of these patients. Many patients live for several days after rupture because the leak is contained by the posterior parietal peritoneum. This author encountered one patient who survived for three months after rupture of an abdominal aortic aneurysm into the retroperitoneal tissues. Free rupture into the peritoneal cavity is rare, but when it occurs, death is usually very rapid.

For these reasons the physician can very often establish the diagnosis in time for treatment.

Pain. The pain is throbbing before rupture but tends to become steady when rupture has occurred. It is located directly over the aneurysm, but also there is almost always pain in the back or loin. *The pain may be present for four or five days before overt rupture and collapse.* After rupture, radiation into the testicles, rectum, or groin in common.

Nausea and vomiting. These are not usually prominent, but they may occur.

Shock and collapse. It is to be hoped that the diagnosis can be established before these occur, but sometimes this is not possible. *Shock or collapse in a patient with a known aneurysm almost always indicates rupture, even if resuscitation can be achieved promptly and fully.*

Hematemesis and melena. Sometimes the aortic aneurysm ruptures into the duodenum, with consequent hematemesis and melena. Surprisingly, such bleeding may start and stop abruptly over a period of many weeks without exsanguination of the patient. Therefore, hematemesis or melena in a patient with a known aneurysm demands active investigation and usually leads to the diagnosis of an aortoduodenal fistula.

Congestive heart failure. On rare occasions, the aneurysm will rupture into the adjacent vena cava, with the sudden onset of severe heart failure that does not respond to the usual medical measures. The latter observation, together with a loud, continuous abdominal murmur, or angiography if necessary, serves to establish the diagnosis.

Abdominal mass. If a preexistent aneurysm is known to be present, the surgeon is greatly aided. Often this is not the case, however, and careful abdominal and flank examinations must be carried out at frequent intervals, since the mass expands rapidly but may be difficult to appreciate in an obese patient.

The mass is usually pulsatile, although sometimes not strikingly so. *It may occupy almost any part of the abdomen* because it usually represents the extravasated hematoma, but the left flank is the most common site.

DIAGNOSIS

The diagnosis of ruptured aneurysm can only be made if the physician strongly considers the possibility. Abdominal and back pain radiating to the groin or testicle in a patient who collapses strongly indicates the possibility of a ruptured aneurysm. *Angiography is contraindicated in such a patient,* and even sending the patient to the X-ray department for a lateral view of the abdomen or an ultrasound scan is extremely dangerous, since irreversible collapse frequently occurs while the patient is away from close and careful supervision. The presence of the pulsatile mass together with the findings described above are sufficient to warrant immediate exploration.

Special intra-abdominal aneurysms

The early history of an *aneurysm of the splenic artery* may closely simulate a gastric or duodenal ulcer, and by pressure on or erosion of the stomach it may cause hematemesis and melena. The pain is felt in the left hypochondrium. Leakage or rupture of an aneurysm of the splenic artery may take place into the lesser sac or into the general peritoneal cavity. When the aneurysm ruptures into the lesser peritoneal sac, the symptoms may be very acute and may closely simulate those of the *perforation of a peptic ulcer or acute pancreatitis;* even acute intestinal obstruction may be suspected. The diagnosis is likely to be made only after exploration of the abdomen.

When the aneurysmal sac ruptures into the general peritoneal cavity, the symptoms will be those of internal hemorrhage, the exact source of which is unlikely to be diagnosable before exploration. A small aneurysm of the celiac axis may cause equivocal symptoms; its rupture may cause collapse, the cause of which may be difficult to determine before exploration. I once explored the abdomen of a male patient who gave a history suggestive of a duodenal ulcer and came under my observation on account of abdominal pain and collapse. The peritoneal cavity was found to be full of blood. The source of the bleeding was not obvious, but there were some superficial lacerations of the spleen and that organ was removed. For two days the patient improved, but then further collapse and death occurred. At the autopsy the cause of the bleeding was found to be a small aneurysm of the celiac axis. Ruptures of aneurysms of the renal artery are usually into the retroperitoneal space, causing the sudden appearance of a tender swelling in the right or left iliac and lumbar regions. There will be some fever, and on the right side the condition may closely simulate an appendix abscess. The suddenness of the appearance of the swelling should put one on guard.

Sudden intraperitoneal hemorrhage

Apart from injury and aneurysm, it is very rare for hemorrhage into the peritoneal cavity to occur in the male. Spontaneous rupture of a malarial, and occasionally of an apparently normal, spleen may take place, and there have been a few instances of profuse bleeding into the peritoneal cavity from a ruptured, gangrenous gallbladder. In the female, ectopic gestation is the most common cause (see Chapter 18). Severe intraperitoneal bleeding may result from the rupture of a small graafian cyst of the ovary; this must be suspected when collapse and signs of intraperitoneal bleeding occur in a young female adolescent in whom an ectopic gestation can be excluded.

Use of contraceptive pills has led to the development of benign adenomas of the liver in young women, the initial symptoms of which may be severe, exsanguinating intraperitoneal hemorrhage. This cause must be considered whether the patient has been on the medication for a short time (up to five or six months) or a prolonged period.

Abdominal apoplexy (sudden, severe intra-abdominal hemorrhage) may result from rupture of a nonaneurysmal vessel. In some instances, fibrinoid degeneration of the affected artery has been found on microscopic examination. There have been reports of involvement of the major gastric arteries, as well as of the splenic, hepatic, inferior mesenteric, and pancreaticoduodenal arteries.

With the increasing use of anticoagulant drugs, the possibility of severe intra-abdominal hemorrhage secondary to hypoprothrombinemia must be considered. Hemorrhage usually occurs into the mesentery or mesocolon, with subsequent rupture of the hematoma into the general peritoneal cavity. The intestine itself becomes so suffused with blood that it may appear to be gangrenous if the surgeon has been unfortunate enough to miss the diagnosis. Hematemesis or melena, especially the latter, is prominent, and shock occurs early. Cramping abdominal pain located centrally is almost invariable. When severe hypoprothrombinemia is present, the diagnosis is readily established and operation avoided.

17. Acute abdominal symptoms in women

Acute conditions occurring during pregnancy and the puerperium

Acute abdominal pain in a pregnant woman is always a source of special anxiety, both from the maternal and the fetal point of view. Exploratory operations are not lightly to be advised on account of the risk of abortion, and one therefore needs to be very sure of the indications of the various acute diseases needing intervention before advising that the abdomen be opened.

The following conditions giving rise to acute abdominal symptoms may occur during pregnancy or the puerperium:

1. Persistent vomiting.
2. Ectopic pregnancy.
3. Retroverted gravid uterus.
4. Threatened abortion.
5. Sepsis following attempts at abortion.
6. Pyelitis.
7. Appendicitis.
8. Torsion of the pedicle of an ovarian cyst or pedunculated fibroid, or torsion of a normal ovary.
9. Degeneration of a fibroid.

10. Spontaneous rupture of the uterus.
11. Pelvic peritonitis or cellulitis.
12. Perforated gastric ulcer.

Persistent vomiting. The morning vomiting of pregnancy usually begins in the second month and continues through the third month. *The nausea of pregnancy is frequently ameliorated by ingestion of food or milk, a circumstance virtually unique to this condition.* As a rule it does not cause any anxiety, but when accompanied by any abnormality in the abdomen, it may give rise to doubt. On several occasions I have known serious doubt to occur in pregnant women who were vomiting and also had a slightly painful swelling in the right inguinal region. It was thought that a strangulated inguinal hernia might be causing the symptoms, but examination showed in each case a uterine fibroid that had risen out of the pelvis with the enlarging uterus and had simulated a hernia by bulging out of the inguinal canal. The vomiting in such cases takes place without the occurrence of any pain in the swelling, and on deep respiration it is possible to feel an intra-abdominal swelling move independently of the abdominal wall. Vomiting and pain in the inguinal region also occur because of traction by the enlarged uterus on the round ligament. Pain is usually more prominent than the vomiting, but of course, no hernia is to be found.

I have also known the vomiting of early pregnancy, when accompanied by slight pelvic pain due to congestion or constipation, to be mistaken for appendicitis, but careful pelvic and abdominal examination, and close attention to the symptom sequence, should exclude this condition.

The excessive vomiting due to the toxemia of pregnancy does not concern us here, but it is necessary to make sure that there is no serious intra-abdominal lesion before diagnosing hyperemesis gravidarum.

Ectopic pregnancy. This is so important and common that the next chapter is devoted to it.

Retroverted gravid uterus. This condition, well termed "the disease of the third and fourth months of pregnancy," may give rise to acute abdominal pain felt chiefly in the hypogastrium, where the distended bladder is situated. The fact of pregnancy, considered along with the presence of a hypogastric swelling (larger than the uterus should be for that time of gestation), the occurrence of pain, nausea, and retention of urine (or maybe dribbling incontinence), would cause one to suspect the condition, which should be easily diagnosed after the urine has been drawn off by a catheter.

Threatened abortion. Threatened abortion should cause no difficulty in diagnosis when pregnancy is known to exist, since the uterine bleeding with the lumbar backache and hypogastric pain, and the absence of evidence of any other local abnormality, should sufficiently determine the condition.

Sepsis following an attempt to produce abortion or in the presence of an intrauterine contraceptive device. It is not unknown for women to try either to produce abortion on themselves or to persuade some lay person to attempt a similar disservice for them. Sepsis may follow these attempts, and when the doctor is called in, it may be difficult to obtain any history of interference. The sepsis may take the form of peritonitis, septicemia, or both. There is generally bleeding from the uterus. In any case, therefore, of bleeding from the uterus in a woman who has had a period of amenorrhea and in whom abdominal pain, vomiting, and fever suddenly present themselves, one must bear in mind the possibility of uterine sepsis. Septicemia and peritonitis in such cases are usually rapid and virulent. The onset is unlike that of any other abdominal condition, and the pain not only may be felt in the hypogastrium but may be referred to the back or down the legs. Careful bimanual examination will determine an enlarged, softened uterus and may exclude any other pelvic condition.

Such sepsis may develop in a woman with an intrauterine contraceptive device in place and may be present simultaneously with an early intrauterine pregnancy. Clinically, the picture is virtually identical to that described in the paragraph above. Sepsis may be present even when the device has not perforated the uterus.

Pyelitis. Pyelitis is a not uncommon complication of pregnancy. It occurs usually about the fourth month of gestation, and its onset may have some relationship to the pressure of the growing uterus upon the ureters, especially the right ureter, for the condition is more common on the right.

The symptoms commonly start with a rigor or feeling of chilliness, and the temperature quickly rises to 103°F or thereabout. At the same time pain is felt in one or the other loin (generally the right) under the costal margin. Pressure at the erector-costal angle produces pain. There may be some frequency of and pain on micturition, but this is not constant. There is as a rule no rigidity of the abdominal muscles. In some cases the patient does not feel ill, while at other times there may be severe malaise. Examination of the urine shows turbidity, albuminuria, and the presence of pus in small quantity. Bacteria, usually *Bact. coli* (*Escherichia coli*), will be detected on microscopic examination. The albumin may not be more than is expected to correspond with the pus.

When pyelitis occurs on the right side, differential diagnosis must be made from appendicitis. This is usually easy, for appendicitis seldom starts with a rigor, infrequently causes a fever as high as 103°F or 104°F, and is often accompanied by local rigidity and not so frequently by any urinary trouble. Examination of the urine should settle the diagnosis.

Appendicitis and perforated gastric ulcer. These are misfortunes that may overtake the pregnant woman, just as they can in any other woman, and they should be diagnosable readily by considering the symptoms carefully. Sometimes when they occur in the puerperium they may be mistaken for the results of puerperal sepsis. The symptoms, however, should be readily interpreted if the possibility of their occurrence is borne in mind.

It is to be emphasized that although the typical symptom sequence of appendicitis occurs in the pregnant woman, the final localization of the pain and tenderness may be confusing, especially later in pregnancy. As the uterus enlarges, the cecum tends to be displaced superiorly. The final position of the appendix, and hence the tenderness, is either high in the right flank or almost midepigastric. The condition is therefore easily confused with acute cholecystitis.

Torsion of an ovarian cyst. In the puerperium also, it is not uncommon for a dermoid or other *ovarian cyst* to become *inflamed* or undergo *torsion*, owing to the contusion and displacement consequent on the labor. There will be acute abdominal pain, fever, vomiting, and a tender hypogastrium, while a rounded swelling will be felt near the uterus but separate from it. A distended bladder should be excluded by catherization and a twisted fibroid by noting the relation to the uterus.

A slightly enlarged but otherwise normal ovary has been known to undergo tension in early pregnancy and cause pain and vomiting so that an ectopic gestation was thought a likely diagnosis until the abdomen had been opened.

Degeneration of a fibroid. Necrosis of a uterine fibroid is particularly prone to occur during pregnancy. The symptoms are pain felt locally in the fibroid, which can be palpated through the abdominal wall, slight fever, and nausea or vomiting. In any pregnant patient who is known to have a fibroid, such symptoms would point to degeneration, and in such cases it is sometimes possible for the fibroid to be enucleated without disturbing the pregnancy.

Rupture of the uterus. Spontaneous rupture of the pregnant uterus is a very rare condition leading to severe shock and signs of internal hemorrhage. For details one must consult an obstetrical textbook.

Pelvic peritonitis. Pelvic peritonitis may ensue after childbirth. Sometimes this may be septic in nature. Any vaginal discharge in the patient or the presence of infection of the baby's conjunctival sacs might give the clue. Hypogastric pain and tenderness, vomiting or nausea, and bilateral tenderness in the uterine fornices will be demonstrable.

Acute abdominal diseases peculiar to women apart from pregnancy

Acute salpingitis. Acute salpingitis is commonly due to infection with the gonococcus. Another frequent cause is infection with the staphylococcus or streptococcus. It can also be caused by *B. coli* (*E. coli*) or the pneumococcus. In recent years, as many cases have been attributable to nongonococcal as to gonococcal organisms. Unfortunately, severe pelvic inflammation has sometimes followed the placement of an intrauterine contraceptive device, even when the device has not been displaced from the uterus. The symptoms and clinical findings are similar to those of infections in the absence of the intrauterine foreign body. Localized abscesses and consequent obstruction of the small intestine may follow perforation by certain types of devices.

Symptoms and signs. The picture is that of an attack of pelvic peritonitis—hypogastric pain, nausea or vomiting, and fever that may reach as high as 103°F. Examination shows tenderness on pressure in both iliac fossae and in the suprapubic region. In some cases the lower abdominal wall is rigid and moves badly on respiration, and there may be considerable distention. Vulval examination may show traces of a gonococcal infection, or a purulent vaginal discharge may be present. Palpation of the lateral fornices may cause pain.

Every new generation of students seems to be taught by the previous generation that severe pain on movement of the cervix is diagnostic of salpingitis. Nothing could be no further from the truth since *any pelvic peritonitis may be associated with severe pain when the cervix is moved in a lateral direction*. An acutely inflamed appendix lying against the broad ligament or having ruptured adjacent to the right tube will cause as much pain when the cervix is moved as a severe acute salpingitis.

Diagnosis. The common and chief difficulty is to distinguish appendicitis from salpingitis (see p. 92).

The symptom sequence is usually more characteristic in appendicitis, and the pain is often more strictly limited to the right side. In salpingitis it is common for the pain to be worse on the left side than on the right—an occurrence rarely seen in the *early* stages of appendicitis.

A vaginal discharge in which gonococci may be detected would be significant, but this does not occur frequently.

It is frequently very difficult, and sometimes almost impossible, to be quite certain whether the appendix or salpinx is primarily at fault. Ultrasound examination may be very useful in showing an enlarged tube, and if doubt still remains, laparoscopy sometimes may be of value.

When pelvic peritonitis follows childbirth it may not show itself for a week or more after the birth.

Pyosalpinx. Many cases of salpingitis quiet down and form a local collection of pus within the fallopian tube—pyosalpinx. The condition is usually bilateral. For a time symptoms may abate and be almost negligible, but sooner or later the infection spreads and an extension of inflammation occurs. There will be the symptoms of pelvic peritonitis, as in the case of salpingitis, but in addition there will be felt a bilateral tender swelling in the pouch of Douglas. The symptoms in a ruptured pyosalpinx are often the more serious since there is frequently secondary infection of the pus sac by organisms other than the gonococcus.

Twisted ovarian cyst. An ovarian cyst with a twisted pedicle gives rise to acute symptoms. Hypogastric pain, vomiting, and the presence of a tender swelling in the lower abdomen are the principal features.

The vomiting comes on almost as soon as the pain, so that there is less likely to be an interval between the initial pain and the vomiting, as is usual in appendicitis. On the other hand, vomiting may be completely absent. The pain is often related to the patient's position, and a certain posture may relieve some of the tension on the twisted pedicle. Most often, the patient prefers to be sitting, while a supine position accentuates the pain. With an ovarian cyst there will be a definite, rounded, tender swelling to be felt either by palpating the hypogastrium or by pelvic examination. If the patient is not seen early, peritonitis with accompanying rigidity may prevent the full outline of the tumor from being felt, and it may be difficult to distinguish from other causes of pelvic peritonitis, for example, acute salpingitis with serious effusion, but the symptoms of a twisted ovarian cyst are usually more acute. Here again, ultrasound and laparoscopy may be diagnostic.

Torsion of the pedicle of hydrosalpinx. In this condition the symptoms are similar to those of an ovarian cyst with a twisted pedicle, though usually slighter in degree—pain, nausea, or vomiting, and a tender, movable swelling in one or the other vaginal fornix. In the presence of even the slightest menstrual irregularity it would be impossible to exclude an unruptured ectopic gestation with certainty.

Endometrioma. Rupture of an ovarian endometrioma may cause hypogastric pain, vomiting, and fever; bimanual examination will reveal a swelling on one or both sides of the pouch of Douglas. This clinical picture might easily be mistaken for tubo-ovarian inflammation or even for pelvic appendicitis, and it would as a rule be difficult to make certain before laparoscopy or exploration. Usually, there is a history characteristic of endometriosis, but this is not invariable.

In every doubtful case of acute abdominal disease in women, it is necessary to pass a catheter to make sure that the bladder is empty before making a final decision.

Ruptured graafian follicle or corpus luteum cyst. These two conditions are essentially indistinguishable, except that a ruptured graafian follicle (ruptured follicular cyst or *Mittelschmerz*) occurs almost exactly midway between menstrual periods, while a ruptured corpus luteum cyst occurs at about the time the menses begin or are expected. Each may produce symptoms of an intensity varying from very mild discomfort to severe pain and profound hemorrhage. Some unusually perceptive women are able to note the time of ovulation almost every month by the onset of the typical and regularly recurrent pain; when this history is present, the diagnosis of a more severe attack is easier.

The onset of pain is usually gradual and the pain is steady in character but will be accentuated by movement or a cough. It is usually felt low in the iliac fossa on either side and occasionally radiates into the groin. Usually there is some degree of anorexia, but nausea and vomiting are uncommon. Tenderness is invariably present in the iliac fossa, but neither tenderness nor rigidity is severe. Bimanual pelvic examination shows definite tenderness of one ovary, which may be slightly enlarged. Occasionally, hemorrhage may be so severe as to cause shock and necessitate operation, particularly if the patient has been using aspirin.

In the milder forms, ruptured graafian follicle or corpus luteum cyst must be distinguished from appendicitis, usually by the absence of the typical sequence of symptoms of appendicitis. Some have claimed that laparoscopy or culdoscopy is useful in these difficult cases, but the cost effectiveness of these methods is yet to be established. Ultrasound may show fluid in the pelvis but, in my experience, rarely shows the cyst since the thin-walled cyst has collapsed as a result of its rupture. If there is severe hemorrhage, it may be nearly impossible to distinguish the two conditions from ectopic gestation (Chapter 18).

18. Early ectopic gestation

By ectopic gestation is meant the development of a fertilized ovum in any place other than the uterine cavity. The rupture of such a gestational sac is a comparatively common occurrence with fairly characteristic symptoms, yet it is often misdiagnosed.

Fertilization of the ovum probably normally takes place in the fallopian tube, and any slight cause may detain the developing ovum, prevent its further progress, and lead to an ectopic gestation. A fertilized ovum has been known very rarely to develop in the substance of the ovary, but this condition is of little practical importance, since one cannot clinically distinguish it. The most common place for ectopic development of the ovum is in the ampullary part of the fallopian tube. More rarely it is found in the isthmial part of the tube, and more rarely still, it may develop in the tubouterine section of the tube or in a rudimentary cornu, which for clinical purposes must be included in the same group (Fig. 32).

In a tubal gestation, growth of the ovum leads to distention of the tube, while the eroding action of the villi leads to a thinning of the wall. Gradual oozing of blood may take place into the peritoneal cavity from the eroded area, or any sudden strain may lead to rupture of the tube. Sometimes the ovum is extruded through the end of the tube into the peritoneal cavity by a process well termed "tubal abortion," or the embryo may die in consequence of hemor-

221

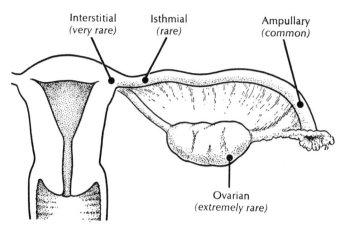

Interstitial   Isthmial   Ampullary
(very rare)   (rare)   (common)

Ovarian
(extremely rare)

Fig. 32. The possible positions of an ectopic pregnancy.

rhage into the sac, or rupture of the sac into the lumen of the tube, and thus may be formed a tubal mole.

If the embryo lives, primary rupture of the sac occurs usually within the first eight weeks of pregnancy, though, in the case of a tubouterine or intestinal gestation sac, rupture need not occur until pregnancy has advanced to the third or fourth month.

Rupture of the sac can cause acute abdominal symptoms, especially those of exsanguinating hemorrhage, but on occasion the symptoms subside with the formation of a mass of blood clots in the pouch of Douglas (pelvic hematocele). It is exceptional for the sac to rupture into the broad ligament.

If the fetus continues to develop after primary rupture of the tube, severe symptoms may be caused at a later date by secondary rupture into the general peritoneal cavity.

Ectopic gestation often results from the first conception of a woman who has been married for some years or in a parous woman who has not been pregnant for several years. In recent years, the use of an intrauterine contraceptive device has been a common antecedent.

Symptoms and diagnosis

A patient with an ectopic gestation may seek advice for abdominal pain:

1. Before the sac has ruptured.
2. At the time of primary rupture of the sac.
3. Some days or even weeks after the rupture.

BEFORE RUPTURE OF THE SAC

If the serious complications ensuing on the rupture of a gestational sac are to be avoided, it is essential that the condition should, if possible, be diagnosed before the tube has given way. Sometimes there are few, if any, symptoms premonitory of rupture, but frequently there are indications of value that enable one to make a diagnosis with sufficient probability.

In a typical case (such as is seldom seen), the symptoms and signs would be:

1. Amenorrhea (for one or two months) or *any* menstrual irregularity.
2. Locally hypogastric tenderness on pressing into the pelvis toward one side.
3. Uterine bleeding.
4. Hypogastric pain.
5. Small, tender swelling in the lateral fornix.
6. The passage of a membrane per vaginam.

Amenorrhea. Though one or two periods may have been missed, the breasts slightly enlarged and full, and even sickness in the morning noted, such definite symptoms are rather exceptional.

There is, however, nearly always some slight irregularity of menstruation, and care must be taken to ascertain the exact particulars of that event. The patient must be asked (1) when was the last period; (2) if that period was before or after the normal time, noting a delay or advancement of as short a period as a day in women who are usually regular to the day; (3) whether the bleeding at the last period was less or more than usual; and (4) whether any slight bleeding has occurred since the last regular period.

Since the gestational sac of an embryo under a month old may rupture, or abort through the end of the tube, it is not absolutely necessary for there to be any irregularity of menstruation. Not uncommonly, the bleeding from the vagina that accompanies tubal abortion is mistaken by the patient for the normal menstrual period, for it may coincide exactly with the expected menstruation. Usually the bleeding is less or greater than normal and is antedated or postdated to the regular period by one or several days. Sometimes what was apparently a normal period ceases for a few days, and then bleeding recommences. All these slight irregularities need to be noted.

Abdominal pain. Abdominal pains of the colicky type felt mainly in the hypogastric or iliac regions may be the chief complaint. Taken in conjunction with the vaginal bleeding, these pains may suggest both to the patient and the doctor a threatening abortion. The pains are probably due to repeated slight intraperitoneal hemorrhage or to the contractions of the fallopian tube.

Uterine bleeding. This has been referred to above. It is not constant. The blood, as it comes away from the vagina, is stated to be darker than the normal menstrual loss, but this is not of much value in diagnosis. What is of greater importance, if it occurs, is the passage of a decidual cast of the interior of the uterus. This seldom occurs until the embryo is dead or aborted into the abdominal cavity, and often occurs after the operation for removal of a ruptured sac has been performed. If it occurs at the time of the early gripping pains, it is of the greatest help in diagnosis. The shreds of membrane do not show any chorionic villi and can thus be distinguished from any embryonal uterine sac.

Hypogastric tenderness. Hypogastric tenderness may be detected on the side of the lesion if the fingers are gently pressed well down behind the pubis. The abdominal wall is not rigid.

Bimanual examination. Bimanual examination reveals a small, rounded, tender, movable swelling to one side of the uterus in one or the other lateral fornix corresponding to the side where the pain and tenderness are elicited. The uterus will also be felt to be slightly enlarged.

To sum up: if the patient and the doctor think that pregnancy has begun, if there is irregular hypogastric pain more on one side than on the other, and if with slight uterine bleeding a tender, rounded, movable lump is felt to one side of a slightly enlarged uterus, ectopic pregnancy should be diagnosed clinically and its presence possibly confirmed by a pregnancy test, for it is only in the ruptured or unaborted stage that the pregnancy test will be positive. On one occasion when on clinical grounds I had diagnosed extrauterine pregnancy and advised operation, the patient went for confirmation of the diagnosis to a hospital. While waiting in the outpatient department of the hospital, she collapsed from internal bleeding and had to be operated on promptly. It is possible, and should be a more frequent occurrence, to diagnose the condition before great internal hemorrhage occurs.

Differential diagnosis. Grandin, in reference to tubal gestation, observes: "The man who suspects every woman of having the condition is the man who is at least liable to err in diagnosis." It is frequently misdiagnosed because it is infrequently considered. An early unruptured tubal pregnancy needs to be distinguished from:

1. Gastritis.
2. Appendicitis.
3. Threatened uterine abortion.
4. Pyosalpinx or hydrosalpinx.
5. Small ovarian or broad ligament cysts.

It is only by the casual observer who has not examined the patient
that the colicky pain of an ectopic pregnancy threatening to rupture
can be mistaken for gastritis. The position and nature of the pain
should direct attention to the pelvis, where examination will soon
provide facts for correct diagnosis.

If the gestation is in the right tube, it is easily mistaken for
appendicitis. Some of the main points in diagnosis may be tabulated:

Symptom or sign	Unruptured inflamed pelvic appendix	Unruptured (but possibly leaking) tubal gestation
MENSTRUATION	Usually regular	Usually some irregularity
UTERINE BLEEDING	Usually none	Usually present
INITIAL PAIN	Epigastric	Hypogastric
FEVER	Slight fever	Usually no fever
VOMITING OR NAUSEA	Present	Unusual
BIMANUAL EXAMINATION	Tenderness but no movable lump	A tender, rounded, movable swelling to one side of the uterus

If the appendix is not situated in the pelvis, there can be little
likelihood of mistake. Neither with a pelvic appendicitis nor with
a tubal gestation is there, as a rule, any rigidity of the abdominal
wall.

A *threatened uterine abortion* is often difficult to distinguish, and
on several occasions I have known uterine curettage performed for
what was thought to be an incomplete abortion but was in reality a
tubal gestation. Bimanual examination ought to distinguish between them, for if it is definitely decided that pregnancy has begun
and there is uterine bleeding, pelvic examination will show in the
one case a slightly enlarged uterus with a small swelling to one side,
and in the other case a larger uterus with no swelling in the lateral
fornix.

The greatest difficulty in diagnosis is in the cast of an intrauterine
pregnancy complicated by a pyosalpinx, hydrosalpinx, or small
ovarian cyst. If severe abdominal pain occurred with any of these

the condition might be impossible to diagnose with certainty, but abdominal section would probably be indicated in any case.

RUPTURE OF A TUBAL GESTATION

Rupture of a tubal gestation is one of the causes of sudden death in young women who have previously been in perfect health. It brings many more to the gates of death.
The symptoms are usually clear:

Sudden abdominal pain.
Vomiting.
Faintness or actual fainting.
Sudden anemia and collapse with a small, rapid pulse and subnormal temperature.
A tender, tumid abdomen.
Free fluid in the abdominal cavity.
Tenderness on pressing the finger against Douglas' pouch.

The pain is sometimes hypogastric, sometimes more general or even epigastric; occasionally pain is felt over the clavicles or even in the supraspinous fossa. Sometimes this clavicular pain may be produced by gently raising the foot end of the bed, thereby allowing any free blood to impinge upon the diaphragm. The anemia should be looked for especially in the lips and tongue and under the fingernails. The sclerotics also have a particularly white appearance. Occasional deep sighing respirations may indicate the severe internal hemorrhage.

The pulse rate is by no means always a good guide to the bleeding, for in some people it takes a very large hemorrhage to cause much increase in rapidity, and in others rapid compensation and restoration of the circulation occur. Nevertheless the pulse is usually rapid and very feeble, and the blood pressure is often considerably lowered.

Examination of the abdomen will show a flaccid but tender abdomen, with some fullness and (exceptionally) slight resistance to palpation in the hypogastrium. Free fluid may be demonstrated, but it is necessary and unwise to do so. The pelvic peritoneum will be tender, and a swelling may possibly be detectable in one lateral fornix. The important points to pay attention to are the sudden onset, the fainting attack, the grave anemia, feeble pulse, and subnormal temperature.

Differential diagnosis. Differential diagnosis must be made from several of the very acute abdominal catastrophes, such as perforation of the stomach, duodenum, or gallbladder, acute intestinal obstruction, acute pancreatitis, acute perforative appendicitis, rupture of a hepatic adenoma with hemorrhage, or torsion of the pedicle of an ovarian cyst. The history of the case, the sudden onset, and the persistence of the extreme pallor and subnormal temperature without rigidity of the abdominal wall in a patient previously well, except for some slight irregularity of menstruation, seldom or never leave room for doubt.

It may be impossible clinically to distinguish rupture of an ectopic pregnancy from *severe hemorrhage from a gaafian follicle* or a corpus luteum cyst. The latter is a rare occurrence and is usually unaccompanied by any of the signs of pregnancy.

CASES WITH SUBACUTE HEMORRHAGE

We must now consider diagnosis after repeated bleedings have caused a hematocele to form.

Many patients come for a diagnosis a short while after an acute rupture of the sac *or after repeated slight hemorrhages have led to the formation of a hematocele* (Figs. 33 and 34). At the time of observation the bleeding may have stopped and the patient may have recovered sufficiently to get about again. The previous serious symptoms may have been attributed by the patient to a simple fainting fit, or the doctor may have thought the trouble due to a cardiac attack. This mistake could only arise in those patients who have had moderate internal hemorrhage, for in the extreme degrees the grave nature of the case cannot be missed.

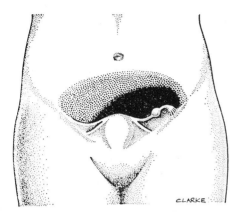

Fig. 33. A hematocele (from the front). (The black area indicates firm and older clot; the dotted portion represents looser and more recent clot.)

After a moderate hemorrhage the slight collapse is quickly recovered from, the pulse may become normal, and the temperature may rise from subnormal to slightly elevated. When there have been repeated slight hemorrhages there may be no history of acute

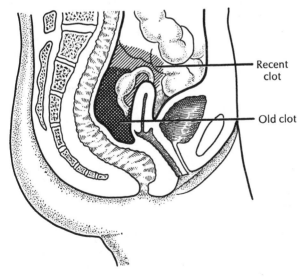

Recent clot

Old clot

Fig. 34. A hematocele (lateral view).

collapse, but usually with repeated attacks of pain the patient will have felt faint. Such a patient comes for advice either for abdominal pain and weakness or for pain and uterine hemorrhage. Occasionally retention of urine may be the reason for seeking advice. Daignosis is made by considering the history of repeated attacks of hypogastric or iliac pain, the irregularity in menstruation, and the conditions found on examination, which are as follows:

1. The patient looks anemic.
2. The temperature is normal or slightly elevated (100°F or 101°F).
3. The pulse is either normal or increased (100 to 120) in frequency.
4. There is a fullness of the lower abdomen (due slightly to distention but chiefly to the blood clot).
5. The lower abdomen is tender on pressure, especially on the side of the tubal gestation.
6. Rigidity of the abdominal wall is absent, though the patient may show resentment of any deep pressure by contracting the muscles.
7. There is usually some uterine bleeding.
8. Pelvic examination reveals a fullness and resistance in one or both lateral fornices. Sometimes a definite harder swelling may be felt. If the blood clot is firm, bimanual examination reveals a definite tumor filling the pouch of Douglas. Pain is always elicited on pressure on the swelling (whether vaginally or per abdomen).
9. There may be retention of urine or frequency of urination.
10. On two occasions I have obtained a positive obturator or thigh-rotation test (see p. 41).

Additional evidence of internal hemorrhage may sometimes be obtained in the form of a bluish or purplish discoloration in the region of the navel (Cullen's sign). Such discoloration, however, is very inconstant.

Vomiting sometimes occurs with the attacks of abdominal pain but is not a constant or important feature. Constipation may be present. Diagnosis is usually clear from the above symptoms and signs. When the pregnancy has advanced two or three months, the softening of the cervix and enlargement of the breasts may help in diagnosis.

Differential diagnosis. Differential diagnosis from pelvic appendicitis, pyosalpinx, or retroverted gravid uterus is sometimes difficult.

Pelvic appendicitis. This may give rise to nearly all the signs and symptoms of a small pelvic hematocele, and the two conditions are frequently confused in diagnosis. Hypogastric pain, vomiting, local tenderness, slight fever, and tenderness on rectal or vaginal examination are commonly present in both conditions. Even the thigh-rotation test may be positive in both cases, and in neither instance should one expect to find abdominal-wall rigidity. The main points of distinction are that in an ectopic pregnancy there is nearly always *some* irregularity of menstruation or even definite amenorrhea for a month or two, usually uterine bleeding, and possibly the symptoms of early pregnancy and the appearance of anemia. In addition, the history of onset is suggestive of internal hemorrhage and is quite different from the usual onset of appendicitis, in which the symptom sequence may be definitely helpful in diagnosis.

Pyosalpinx. With a pyosalpinx there should not be menstrual irregularity, though menorrhagia is not infrequent, and there should be a history of previous pelvic inflammation or of leukorrhea, endometritis, or definite gonococcal infection. The uterine cervix will not be softened in pyosalpinx, and pain on micturition and retention of urine are less likely to occur.

Retroverted gravid uterus. This may be confused with an ectopic pregnancy that has gone on to the third or fourth month of gestation. Both may cause retention of urine. But if the bladder is emptied by catheter, it should be possible to make out the determining factor in diagnosis (i.e., the absence or presence of the uterine fundus in the normal position). The bleeding from a retroverted gravid uterus would be that of fresh blood, indicating threatening abortion; the blood lost in an ectopic gestation would usually be darker. Anemia would be observed only in the ectopic pregnancy.

If the fetus of an ectopic pregnancy survives the rupture until the later months of gestation, the condition will cause the formation of an abdominal tumor in which the fetal parts may be felt easily under the abdominal wall; at the same time, the symptoms of pregnancy will be very evident. It is possible that acute symptoms might occur at this stage, but this is rare and need not be considered here.

19. Acute abdominal disease with genitourinary symptoms

Acute abdominal disease with genitourinary symptoms may be due to primary disease of the genitourinary organs or to secondary irritation of those organs consequent on disease of other viscera. The main symptoms that may call attention are:

1. Painful swelling in the loin.
2. Renal colic.
3. Disorders of micturition—pain, frequency, retention.
4. Abnormalities of urine—hematuria, albuminuria, porphyrinuria.
5. Pain in the testicle or along the spermatic cord.
6. Tenderness at the right erector-costal angle.

PAINFUL SWELLINGS IN THE LOIN

A pyonephrosis or a hydronephrosis may give rise to acute symptoms. In each case there will be found a rounded, tender swelling in the loin. The tumor will be felt to fill out the posterior lumbar region. The colon may be felt to the front and inner aspect of the swelling, but if the loin is well filled by the tumor, that alone is usually quite sufficient to diagnose a renal swelling. In both pyonephrosis and hydronephrosis there should be a history of urinary trouble. Acute attacks of pain are prone to occur in a recurring hydronephrosis, and gastrointestinal symptoms (flatulence and vomiting) may accompany the attacks. If seen during an attack, however, the swelling is characteristic. In a pyonephrosis there are usually the signs of toxic absorption—fever, furred tongue, and a toxemic appearance. With a hydronephrosis the symptoms are less severe. Sometimes it is difficult to obtain a history of renal symptoms, and then it may be impossible by clinical examination alone to distinguish a pyonephrosis with accompanying perinephritis from other causes of local suppuration, such as appendicitis or diverticulitis—particularly in an obese patient.

Polycystic disease of the kidneys may give rise to uremia with vomiting and abdominal distention. *This might be mistaken for intestinal obstruction,* but the presence of tumors in both loins would make one suspicious, and the occurrence of albuminuria, high blood pressure, and a high blood urea nitrogen would make the diagnosis certain. A polycystic kidney usually has an irregular surface due to the smooth cysts, which vary considerably in size.

RENAL COLIC

Pain of the renal colicky type may be caused by:

1. Stone in the pelvis of the kidney.
2. Stone in the ureter.
3. Blood clot or uratic debris in the ureter.
4. Appendicitis.

In renal colic the pain starts in the loin and frequently, but not always, radiates to the groin or to the corresponding testicle. It may be due to anything solid or semisolid passing down the ureter. There is usually no difficulty in diagnosis, though occasionally appendicular colic or even appendicitis may cause pain of a similar nature. But in appendicitis severe enough to cause the simulation of renal colic there will usually be persistent local muscular rigidity, which is not usually found in renal colic (see Chapter 12).

DISORDERS OF MICTURITION

Pain on or frequency of urination is found in nearly all cases of pyelitis and in many cases of renal colic, but it is often also noted in appendicitis and other causes of pelvic peritonitis. Examination of the urine should prove or disprove a pyelitis.

In appendicitis, pain on or frequency of urination may be due to irritation of the renal pelvis, ureter, or bladder by the inflamed appendix or contiguous peritonitis. When this symptom is accompanied by pelvic tenderness, or a tender lump is felt per rectum, or a positive obturator test is obtained, the appendix will usually be found in the pelvis irritating the bladder. This set of symptoms is found whenever an inflammatory mass lies adjacent to the bladder. Thus, sigmoid diverticulitis or Crohn's disease may cause urinary frequency, urgency, fever, and a pelvic mass. These two conditions must always be considered under these circumstances. When unaccompanied by the other signs, the appendix is generally either near the kidney, where it may irritate the renal pelvis, or at the brim close to the right ureter.

In pelvic hematocele due to a ruptured ectopic gestation there are frequently urinary symptoms. Sometimes there is retention of urine, sometimes slight pain or frequency. In a very anemic woman with abdominal pain and urinary symptoms, it is well to think of ectopic gestation. Acute distention of the bladder may cause very severe hypogastric pain and sometimes pain in the lower lumber region; there should be no difficulty in palpating or percussing out the distended bladder, and one must not be misled by the fact that urine may be passed by the process of overflow incontinence. In those rarer cases in which the bladder does not distend upward but rather backward and upward it may be difficult to make out its position by percussion, but bimanual examination will determine it, and the passing of a catheter will help the diagnosis.

ABNORMALITIES OF URINE

Hematuria frequently follows renal colic and may sometimes enable one to trace to its source a renal pain that was not quite typical.

Albuminuria is an exceedingly important finding, for uremic symptoms may very closely simulate intestinal obstruction, and it may only be by the discovery of albuminuria that one is put on the right track. Many patients with nephritis have undergone needless and harmful abdominal section because of neglect to test the urine. Toxic albuminuria is found in many septic states of the abdomen, but it is usually not difficult to judge when the toxemia is severe enough to produce that symptom. In attacks of acute porphyria, the urine turns brown or reddish-brown on standing.

TESTICULAR PAIN

Pain in the testicle is met with in renal colic and in a few cases of appendicitis. In the latter instance it may be due to irritation of the sympathetic branch that accompanies the spermatic artery, or it may be a true referred segmental pain across the tenth spinal segment. Ruptured aortic aneurysm with retroperitoneal hemorrhage often causes unrelenting testicular and groin pain.

Torsion of the imperfectly descended testicle may cause extreme pain in the inguinal region and may stimulate a strangulated inguinal hernia. Vomiting occurs and shock may be severe. The pain has the usual sickening character of testicular pain. The absence of the testicle from the same side of the scrotum will be noted and will make diagnosis easy.

Thrombosis or suppurative phlebitis of the veins of the spermatic cord cause pain in the inguinal region and sometimes in the iliac region of the abdomen. But the swelling and the painful area extend right down to the testicle, which becomes swollen and painful.

Retraction of the testicle is occasionally noted in cases of appendicitis. It is due to reflex contraction of the cremaster muscle.

TENDERNESS AT THE ERECTOR-COSTAL ANGLE

Tenderness on pressure at the right erector-costal angle (i.e., the usual position for eliciting tenderness of the right kidney) is noted in many cases of appendicitis, especially when the appendix is retrocecal in position.

Renal symptoms may be caused by any retroperitoneal lesion in the region of the renal pelvis; thus a duodenal ulcer leaking posteriorly or a retroperitoneal perforation of the common bile duct may cause frequency of micturition or even slight hematuria.

Pancreatitis must never be forgotten as an important cause of left flank pain and tenderness.

20. The diagnosis of acute peritonitis

Apart from hemorrhage, the cause of death in nearly all fatal acute abdominal cases is either secondary peritonitis or acute paralytic or mechanical obstruction of the intestines. Spreading or general peritonitis is the most common single cause of death. Most cases, unless the patient is actually moribund, sooner or later demand an opening of the abdomen. At operation fluid may easily be obtained for examination, but even before operation it is possible to obtain some by puncturing the abdomen in the midline midway between the navel and the pubis by means of a fine-bore needle. The exudate is stained for organisms and examined microscopically.

Infective organisms may reach the peritoneum:

1. Through a wound of the abdominal wall.
2. Via the bloodstream.
3. From the viscera within the abdomen (most common).
4. Through the diaphragm (almost never) or by lymphatic extension from the thigh (very uncommon).

The only commonly blood-borne organism of importance is the pneumococcus, which may cause severe so-called primary peritonitis.

Organisms may reach the peritoneum from the contained viscera either by (1) rupture of a viscus or by (2) escape through the diseased wall of any viscus. In the female there is the additional path of infection via the fallopian tubes. Frequently a local abscess may form either extra- or intraperitoneally near a diseased viscus, and later this abscess may burst into the general peritoneal cavity or into an adjacent viscus.

CAUSES

The common causes of general peritonitis are the conditions already described in previous chapters. They comprise disease or rupture of the hollow viscera and ruptured abscess of the solid viscera.

Perforation of:
   Appendix vermiformis.
   Gastric or duodenal ulcer.
   Typhoid or tuberculous ulcer of the small intestine.
   Dysenteric or stercoral ulcer or of a diverticulum of the colon.
   Gallbladder or biliary ducts.

Gangrene of:
   Strangled coil of gut.
   Intussuception.
   Volvulus.

Infection spreading from:
   Pyosalpinx.
   Infected uterus.
   Pyonephrosis.

Rupture of:
   Liver abscess.
   Splenic abscess.

A less acute form of peritonitis may result from the escape of sterile bile or urine into the peritoneal cavity.

SYMPTOMS

The symptoms of peritonitis vary greatly according to the part and extent of the peritoneum involved, the nature of the infective agent, and the acuteness of onset. Help in diagnosis will be obtained by regarding the symptoms as being roughly grouped into two classes, reflex and toxic:

Reflex
$$\begin{cases} \text{Pain.} \\ \text{Vomiting.} \\ \text{Muscular rigidity.} \\ \text{Anxious facial expression.} \end{cases}$$

Toxic
$$\begin{cases} \text{Alteration of temperature. Collapse.} \\ \text{Distention. Intestinal paresis.} \\ \text{General toxemia.} \end{cases}$$

REFLEX SYMPTOMS AND SIGNS

The early and reflex symptoms of peritonitis may, in the absence of initial collapse, be extremely equivocal. They are more likely to be definite in young persons; conversely, they may be insignificant in old and debilitated patients.

Pain. Pain is the most constant symptom. It may be confined to the local area of inflammation or referred more generally over the abdomen. As the peritonitis extends the pain area also extends, though the maximum pain is nearly always felt at the initial focus. Tenderness is a constant feature over any focus of peritoneal inflammation. Even when rigidity is absent, pain is usually felt on pressure over the affected site. Only rarely have I seen this tenderness absent—either as the result of extreme toxemia dulling the sensorium or as the result of extreme muscular rigidity that apparently prevented transmission of the applied pressure to the underlying inflamed area.

Vomiting. Vomiting is common at the onset of peritonitis but is usually infrequent until late in the case. The later vomiting is usually obstructive in character.

Rigidity. Muscular rigidity is a common accompaniment of the early stages of peritonitis, *but only when the part of the peritoneum affected lies in the demonstrative area* (i.e., the anterior and lateral abdomen). It is best seen in perforation of a duodenal or gastric ulcer, whereby a large part of the demonstrative section of the peritoneum is irritated. It is generally absent or only slightly demonstrable when the peritonitis is limited to the pelvis. In those cases in which rigidity is present in the early stages the muscles relax as the peritonitis progresses, until in the final stage it is almost absent. Rigidity is also either absent or difficult to detect in obese people with flabby muscles and in old and weak patients. Rigidity may be of slight degree in some cases of pneumococcal peritonitis and in some of the slowly advancing infections due to *Bact. coli* (*Escherichia coli*) and streptococci.

When sterile bile, urine, or blood leaks into the peritoneal cavity, the resulting irritation may not be accompanied by rigidity of the abdominal muscular wall. In cases of biliary extravasation, several pints of bile may accumulate within the peritoneum before the increasing tumidity and distention, and the presence of free fluid within the abdomen, reveal the serious nature of the condition.

The change in facial appearance, though a guide to the experienced, is something that cannot be indicated satisfactorily in words. The white, drawn face of the patient in initial collapse is by no means constant. In any case, the reaction from initial collapse is so rapid and complete that the facial aspect commonly becomes and remains almost normal for a time. But there is usually an indefinable yet definite aspect of countenance in many persons who may present an otherwise doubtful picture of peritonitis. The Hippocratic facies present in late peritonitis is merely that of extreme collapse.

TOXIC SYMPTOMS

The toxic symptoms of peritonitis are later in appearance and indicate a more serious stage of the disease. An occasional intermittence of the pulse is often one of the indications of advancing peritonitis. As the infection involves the various coils of intestine they become paralyzed and distended, while the intestinal contents stagnate and increasing obstruction results. As the gut becomes paralyzed, peristalsis ceases and no sounds will be heard on auscultating the abdomen. Fluid accumulates in the abdomen at the expense of the circulating blood, and secondary collapse results. At this stage true obstructive vomiting is commonly a feature, and there is noted the small, running pulse commonly described as indicative of peritonitis. Such a pulse is felt in the *latest* stages of peritonitis, and not in the earlier stages when diagnosis is so important.

DIFFERENTIAL DIAGNOSIS

It is not difficult to diagnose a flagrant case of peritonitis, for the pain, vomiting, local tenderness, and muscular rigidity with fever sufficiently indicate the condition; but mistakes are likely to be made either because the symptoms are too slight or because they are atypical. The early symptoms are slight and deceptive when the part primarily affected lies in the pelvis or in some other relatively silent area of the abdomen; they are often atypical in patients who are old, debilitated, or very obese. In the late stages of peritonitis, it is frequently impossible to differentiate this condition from intestinal obstruction, which develops as a late consequence of peritonitis. We would take one more opportunity to emphasize that the condition of the pulse is no true guide in diagnosing early peritonitis.

The conditions that may stimulate peritonitis are:

1. Pleuropneumonia.
2. The colics.
3. Intestinal obstruction.

4. Internal hemorrhage.
5. Some nervous conditions (e.g., tabes, hysteria).

The differentiating points are enumerated in other parts of this book.

Primary pneumococcal peritonitis

To diagnose primary peritonitis without positive proof is dangerous. Such proof is established only by demonstration of the pneumococcus by stain and culture. The surgeon who explores a patient with peritonitis caused by mixed organisms but cannot find a clear cause for the infection and thus assigns the diagnosis of primary peritonitis has usually failed to search adequately for the cause.

By primary pneumococcal peritonitis is meant that form of infection with the pneumococcus in which the peritoneal symptoms are the predominating feature of the illness. It is reasonable to suppose that the peritoneal lesion is nearly always secondary to some other focus in the body, but when that focus is latent or of minor importance it is customary to use the term above.

Blood infection without any ascertainable focus of origin does occur, but more frequently infection spreads from the fallopian tubes or possibly from the intestine. The disease is much more common in females under the age of ten. In tropical Africa, pneumococcal peritonitis is more common in adult women.

The symptoms vary considerably. They are those of peritonitis of varying grades of severity. One can with advantage describe three stages to the disease—the stage of onset, an intermediate stage, and a residual stage.

ONSET

The stage of onset starts abruptly with abdominal pain, vomiting, and fever, which may reach 104°F or 105°F. In some instances there is noticeable diarrhea. Frequently there are toxic symptoms out of proportion to the local findings, and delirium may be noted. The abdominal pain may be diffuse but is usually more definite in the hypogastrium. The abdominal wall may be rigid but more often is soft, and there may be slight local tenderness. In the very severe cases, which are always septicemic for a time, death may occur within two or three days unless the condition is recognized and promptly treated.

INTERMEDIATE STAGE

The intermediate stage corresponds to the formation of pus within the abdominal cavity. The toxic symptoms lessen, but irregular fever persists, and there is a gradual tumesence of the abdomen. Leukocytosis will be observed. Abdominal pain in this stage is very slight, and the continued fever and a painless, tumid abdomen may cause a suspicion of tuberculosis.

RESIDUAL STAGE

The residual stage corresponds to the definite localization of pus and the absence of all acute symtoms except fever of an irregular type. The pus may be in the pelvis, or in the central part of the abdomen, or even in the subphrenic region.

DIAGNOSIS

The differentiation in the first stage has to be made from acute appendicitis. This is sometimes very difficult, but if the initial fever is high, if the abdominal signs are rather indefinite, and especially if there is slight delirium and troublesome diarrhea, one should suspect pneumococcal peritonitis. The condition occurs chiefly in young girls, and some help may be obtained by examining the vaginal secretion for the pneumococcus, since the fallopian tubes are the common route of infection. If there is any serious doubt it is wiser to recommend exploration, since more harm is likely to result from allowing a septic appendix to perforate than from exploring an early pneumococcal peritonitis.

In the intermediate stage the continued fever, tumidity of the abdomen, and comparative absence of abdominal pain and tenderness may lead to a suspicion of tuberculous peritonitis or even typhoid fever. The history of onset, the palpation of any enlarged glands or masses in the abdomen, or the positive result of a test for tuberculosis may help to diagnose the former, while the absence of leucocytosis, an enlarged spleen, and a positive agglutination test would point to the latter.

In the third or residual stage, a definite swelling will be detected in some part of the abdomen; this is a residual abscess and may be in the pelvis or upper or middle abdomen. The diagnosis in this stage has to be made from the other causes of abdominal tumor and abscess. The history of the previous illness will be of the utmost importance in deciding on a diagnosis.

It has been mentioned above that pneumococcal peritonitis may not give rise to the muscular rigidity that is usual with most other forms of acute peritonitis. I have known the combination of abdominal pain, vomiting, and lax abdomen to give grounds for diagnosing acute intestinal obstruction.

If there is coincident pleuritis or pericarditis, the symptoms of peritonitis may be entirely overshadowed. In the early stages of a pneumococcal polyserositis it is often very difficult to determine the exact diagnosis, but if definite signs declare themselves in the lung, it would be inadvisable to recommend opening the abdomen until there are sufficient grounds for suspecting the presence of pus. If the thoracic condition has been successfully treated but the patient does not progress, special care must be paid to the abdomen, for the absence of rigidity and the comparatively slight pain may allow a considerable collection of pus to go unobserved.

The prognosis in cases of pneumococcal peritonitis has been greatly altered for the better by the fact that the organism is sensitive to some of the antibiotics. Early cases can usually be successfully treated by administration of penicillin, and thereby operation may be altogether avoided.

Primary peritonitis in cirrhosis of the liver

Recently it has been noted that patients with cirrhosis of the liver, especially the alcoholic form, are particularly susceptible to the development of a peculiar type of primary peritonitis. All reported cases have had rather profound preexisting ascites. It is an atypical feature of this type of peritonitis that pain, tenderness, and rigidity are extremely mild and, on occasion, even absent. Fever and leukocytosis are similarly low grade. The diagnosis can only be made by finding organisms and leukocytosis in the ascitic fluid of a cirrhotic patient who has some mild abdominal signs and symptoms like those described above together with fever.

21. The early diagnosis of abdominal injuries

Every obviously serious case of abdominal injury is nowadays referred to the surgeon; but since there are many cases in which, though serious damage has resulted from the injury, the symptoms are not obvious and indeed may remain latent for some hours, it is useful to summarize the main points in diagnosis.

We shall confine ourselves to the consideration of those injuries in which there is no open wound of the parietes, and of those penetrating wounds inflicted by a knife or stiletto, for when a gunshot or shrapnel wound of the abdominal wall exists, there should be no question that the treatment is surgical.

Nonpenetrating injuries of the abdomen are most commonly due to severe crushes, as would follow from a heavy vehicle running over the abdomen. Automobile accidents are also a common cause of intra-abdominal injuries. Another type of violence is the sharp, circumscibed blow due to a kick, punch, or the sudden impinging of any hard body against the abdominal wall. Severe injury may also follow the strain on the visceral attachments consequent on a fall from a height, or a sudden trip-up causing a violent fall forward. When violence is applied against the abdominal wall, the kind of injury produced depends to a certain extent upon the preparedness of the patient for the blow and the consequent rigidity or flaccidity of the abdominal muscles. If the muscle is taken unawares more serious intra-abdominal mischief is to be expected, whereas if the muscle is rigid it may mitigate or prevent injury of the underlying viscera.

Any injury of the abdominal wall, however slight, may be accompanied by serious lesions of the viscera, and the latter may be seriously injured without any visible sign of injury to the abdominal wall. When there is no external wound the thought of visceral injury may not be present in the mind of the observer, and grave lesions may be allowed to progress considerably before they are suspected.

The solid viscera of the abdomen (liver, spleen, pancreas, kidneys) are situated high up in the abdomen, largely under the cover of the ribs; the hollow tubes (intestines, bladder, ureter, and stomach) are more exposed. Despite this, the solid viscera are more commonly injured than the hollow ones, with the exception of the bladder, which may be compared to a small organ when it is filled with urine, a condition in which it is more likely to be ruptured than when empty. Solid viscera rupture more frequently than hollow ones because they are generally more fixed and lack compressibility, and are therefore more easily impinged against adjacent bony structures or torn from their attachments.

Injury to solid viscera causes hemorrhage, and injury to the hollow viscera usually causes peritonitis. Both types of lesion may be accompanied by shock.

SHOCK AND HEMORRHAGE

Shock is shown by pallor, feeble pulse, sweating, slow shallow respiration, and cold extremities, but unless there is some serious lesion the symptoms may soon subside. If the state of shock lasts longer and seems out of proportion to the evidence of intra-abdominal injury, examination may reveal a pneumothorax or other chest lesion. Injuries of the upper abdomen cause more serious shock than those of the hypogastric region. Renal contusions also cause severe shock. The observer should recollect that in a very shocked patient the reflexes are diminished; consequently, rigidity of the abdominal wall may be absent even in the presence of peritoneal irritation.

Because complete hemodilution after a hemorrhage requires at least eighteen hours and because capillary refilling of the circulation may take four or five hours to produce a significant drop in the hematocrit or hemoglobin concentration, *a normal hematocrit within the first few hours should not be misconstrued to indicate absence of intra-abdominal hemorrhage.*

DIAGNOSIS OF ABDOMINAL INJURIES

The essential point in diagnosis is to estimate the different proportions that shock, hemorrhage, and peritonitis take in the production of the observed symptoms and to judge from this the viscus injured and the nature of the lesion.

It is frequently impossible to give a definite opinion for a few hours after the accident. *Shock out of proportion to visible injuries almost always means serious intra-abdominal hemorrhage if the visible injury can be excluded as a cause.* Initial shock usually subsides within two or three hours, and then symptoms of hemorrhage or peritonitis become increasingly evident. If shock is still present after three hours, there is nearly always some serious visceral lesion. Pain, vomiting, local or general muscular rigidity, tenderness, alteration of pulse rate, shallow respiration, diminution of liver dullness, free fluid in the abdomen, and silence on auscultation of the abdomen—these are the main symptoms to note. By taking into consideration the part of the abdomen struck, it is possible in many cases to say which viscus in injured and what is the nature of the injury.

The unconscious patient with abdominal injury. Automobile accidents may cause simultaneous injuries to the head and thorax or abdomen, and the practitioner may have to diagnose the nature of an abdominal injury in an unconscious patient.

Shock in an unconscious patient should never be attributed to the intracranial injury. The cause of the hemorrhage will almost

*invariably be found in the thorax or abdomen in such a patient
if no overt major fractures or overt external hemorrhage is present.*
Often physical signs may show the need for exploration, but
when in doubt it may be advisable to perform lavage of the abdominal cavity. The examination of the fluid thus obtained will
demonstrate whether operative intervention is needed.

Hemorrhage. Due to the anatomical disposition of the parts, hemorrhage usually follows lesions of the upper zone in which lie the
liver and spleen. The main abdominal vessels are more likely to be
injured by violence directed against the central portion of the
abdomen. When the liver or spleen is severely torn, the symptoms
of shock and hemorrhage are extreme, and death frequently follows
soon after the injury. In lesser degrees of injury of the solid viscera
(or of the mesenteric blood vessels), evidence of bleeding gradually
appears. Increasing restlessness and pain, progressive pallor of the
lips and fingernails, a rising pulse rate, a decreasing blood pressure
and urinary output, and the demonstration of movable dullness in
the flanks are sufficiently indicative. Occasionally the symptoms of
hemorrhage may abate for a day or two, and then become alarm-
ingly evident after some exertion (e.g., straining on the bedpan).
This is more likely to occur after injuries to the spleen. The pulse
rate is a good but by no means infallible guide in abdominal
hemorrhage. It may remain fairly slow (not above 100) until the
abdomen is full of blood and then suddenly bound up to a rapid rate
as the blood pressure falls. Presumably the increase in pulse rate
takes place when the cardiovascular compensating mechanism
fails. The pulse rate should be taken every half hour in cases of
suspected hemorrhage, and if injury of the spleen is suspected, a
search should be made for local tenderness. An X-ray of the splenic
region might show a fractured rib or some abnormal appearance
just above or below the diaphragm.
The significance of pain on top of the shoulder due to irritation of
the diaphragm has been referred to in an earlier chapter. It is well

known that rupture of the spleen may be indicated by pain on top of the left shoulder. Similarly shoulder pain may follow a wound of the liver. John Hunter noted this in his account of gunshot injuries: "From a wound in the liver there will be pain in the part of a sickly or depressing kind; and if it is in the right lobe there will be a delusive pain in the right shoulder, or in the left shoulder from a wound in the left lobe."

Contusion or rupture of the kidney is frequently accompanied by very severe and alarming shock, when generally passes off within an hour or two (e.g., the "kidney punch" in boxing). The later symptoms depend on the extent of the injury and the condition of the renal capsule. In slight cases where the capsule remains intact, hematuria, local tenderness, and sometimes renal colic due to the passage of clots down the ureter comprise all the symptoms. If the renal capsule is torn a retroperitoneal hematoma is formed, and sometimes urine is extravasated into the tissue behind and around the kidney; this leads to cellulitis accompanied by malaise, irregular fever, local swelling, tenderness, and muscular resistance. Associated adynamic ileus is very common. An abscess may result. If the urine is infected, the symptoms are more acute. Should the peritoneal covering of the kidney be torn, symptoms of intraperitoneal hemorrhage will result. Most cases of hematuria resulting from renal contusion stop spontaneously, but a rising pulse rate, accompanied by continuing hematuria or by an increasing retroperitoneal perirenal swelling, calls urgently for exploration. The vast majority of blunt injuries to the kidney are safely treated nonoperatively; only those in which continuing hemorrhage occurs require operation.

PERITONITIS

Peritonitis is usually the result of rupture of a hollow viscus. The intestines and bladder are most commonly injured, the stomach, gallbladder, and ureters rarely. Sometimes the injury is only sufficient to bruise the walls of the viscus. If the stomach is thus contused, the result is vomiting and sometimes hematemesis; if the colon is bruised, the passage of blood per anum and diarrhea due to traumatic colitis may follow.

Bruising of the bladder causes slight hematuria, but if the contusion is severe there may be vomiting and local muscular rigidity, even though no rupture has occurred.

RUPTURE OF INTESTINE

This is the most common cause of peritonitis after abdominal injury; it is a condition fraught with danger of almost certain death if not diagnosed early, yet the signs and symptoms are often equivocal for some hours. Generally the tear only involves a portion of the circumference of the gut, but occasionally a complete severance is caused and a gap is left between the two ends that may be temporarily closed by contraction of the involuntary intestinal muscle. When intestine is injured, its peristaltic movements stop due to a reflex or direct paresis of its walls. If the rent is small, the edges of the mucous membrane pout and fill the small gap; a small amount of intestinal contents escapes and sets up a local plastic peritonitis that glues together the coils of intestine. The patient takes nothing by mouth, and the gut remains at rest. A lull in the symptoms gives a false sense of security. After a few hours, when the observer may have decided that there is no serious intra-abdominal lesion, food is taken, the intestines are excited to peristaltic contraction, and the opening in the gut is unsealed. Peritonitis then develops more or less rapidly according to the size of the opening and the number of adhesions.

The important earlier indications of peritonitis are:

1. Pain.
2. Local tenderness.
3. Local muscular rigidity.
4. Vomiting.
5. Shallow abdominal respiration.

The later evidence of peritoneal infection includes:

1. Elevation of pulse rate and temperature.
2. Increasing distention.
3. Tenderness of the pelvic peritoneum.
4. Movable dullness in the flanks.
5. Obliteration or diminution of liver dullness (caused by gas in front of the liver).

The patient often has an anxious facial expression and may show unusual restlessness. In injuries of the upper abdomen importance must be attached to the shifting of the pain to the hypogastrium, due to the inflammation caused by the escaped intestinal contents that gravitate to the pelvis. *The presence of normal intestinal sounds does not exclude the possibility of a serious intra-abdominal injury.*

A plain X-ray film of the abdomen may be useful by showing free gas.

The prognosis in cases of ruptured intestine is very bad unless a diagnosis is made and operation undertaken soon after the injury. Hence the need for early diagnosis.

Provided that there is no lesion in the chest and that renal trauma can be excluded, one is probably dealing with a case of ruptured intestine in the following conditions:

1. When severe abdominal pain persists for more than about six hours after an injury and is accompanied by any of the following:

    (i) vomiting, especially bilious vomiting;
    (ii) a pulse and temperature gradually rising from normal;
    (iii) persistent local rigidity that tends to extend;
    (iv) deep local tenderness with shallow respiration.

2. When abdominal pain is absent or very slight and anemia is not increasing, but the pulse rises steadily hour by hour and the patient is very restless or listless.

When marked diminution of liver dullness occurs with any of the above symptoms, or if there are signs of free fluid in the abdomen, or if rectal examination shows the pelvic peritoneum to be very tender, the indications for operation are imperative.

It is assumed that opening the abdomen would be advised without any delay if the symptoms of peritonitis are typical.

RETROPERITONEAL RUPTURE OF THE DUODENUM

It is possible for the duodenum and parts of the colon to be ruptured behind the peritoneum. The symptoms are then due to a retroperitoneal cellulitis, with some inflammation of the contiguous peritoneum, namely, local pain and muscular rigidity, shallow respiration, vomiting, and rise in pulse rate and temperature. In a case of retroperitoneal injury to the duodenum that was under my care, the pain at first was very slight but became greater hour by hour until I felt constrained to open the abdomen though there were few local signs to guide me. In fact, this is the usual sequence of events. So slight may the initial pain, tenderness, and fever be that the diagnosis may be delayed for several days. Retroperitoneal gas demonstrated by plain X-ray is diagnostic but is not always present. If the patient's condition permits, the oral administration of a water-soluble dye may demonstrate the rent. Whenever exploring the abdomen for blunt trauma, the retroperitoneal duodenum should always be examined. The slightest discoloration or ecchymosis usually means rupture.

RUPTURE OF THE URINARY BLADDER

This usually occurs in connection with a fractured pelvis but may result from a blow on the hypogastrium when the viscus is distended. In children the bladder is situated higher up in the abdomen and is therefore more liable to injury. The symptoms vary according to whether the rent is intra- or extraperitoneal. If within the peritoneal cavity symptoms of pertonitis ensue, it must be remembered that *sterile* urine does not at first cause a very acute inflammatory reaction. There may be hematuria. Rupture outside the peritoneum tends to cause extravasation of urine and consequent cellulitis in the suprapubic and perineal regions.

In cases of suspected rupture of the bladder two additional means of diagnosis are available:

1. The bladder is emptied by a catheter, and a measured quantity of saline solution is introduced and again drawn off. Any serious discrepancy between the amounts put in and drawn out suggests rupture of the bladder.
2. More certain evidence of a ruptured bladder may be obtained by introducing into the bladder a suitable amount (200 or 300 ml) of radiopaque solution and taking an X-ray photograph. The opaque solution will readily be seen outside the bladder if proper anteroposterior and bilateral oblique films are taken.

In a patient with a fractured pelvis the membranous urethra may be torn completely across, and the neck of the bladder with the torn portion of attached urethra may retract from the triangular ligament. In such cases the bladder sphincter may remain contracted and the viscus may become overdistended. A tense, tender swelling will thus be detectable in the hypogastrium. The tenseness and tenderness of the lower abdomen thus produced may cause the diagnosis of peritonitis to be made erroneously. In such cases passage of a urethral catheter *should not be attempted* if a drop of blood is seen at the tip of the penis, since a partial tear of the urethra may be converted to a complete one by forceful catheterization. Rather, a urethrogram should be done. Rectal examination shows that the prostate is dislocated superiorly and that tenderness is severe.

RUPTURE OF THE STOMACH AND PANCREAS

This quickly leads to symptoms of general peritonitis; it is generally accompanied by some other lesion such as injury to the spleen or liver. The part of the stomach likely to be ruptured by injury is the part that is seldom the site of ruptured ulcer (i.e., the greater curvature). The gas that escapes may for a time be localized and form an area of superficial tympanitic resonance on percussion. The symptoms of pancreatic injury are in no way distinctive. The most severe injuries may be present with surprisingly few symptoms or abdominal findings. Increasing epigastric pain tends to spread lower down the abdomen. Initially, the pain and tenderness may be surprisingly slight. Evidence of hemorrhage is usually present, but it may not be pronounced. The serum amylase is usually but not always elevated. The pancreas is usually found to be torn across the body or neck in the region of the superior mesenteric vessels despite the general paucity of clinical findings. At operation, a large hematoma at the root of the transverse mesocolon, if present, should always be explored since it usually indicates transection of the pancreas. The condition is usually, but not always, associated with other injuries.

DIFFERENTIAL DIAGNOSIS

It is very important always to *examine the thorax carefully* for pneumothorax or hemopneumothorax. Symptoms very suggestive of serious abdominal injury may be produced by either of these lesions. Abdominal pain and rigidity and the general signs of shock and hemorrhage may follow a fracture of a rib with rupture of a lung and the consequent presence of blood and air in the pleural cavity. Hence there is a need for careful thoracic examination. In cases of doubt, radiography of the chest by means of a portable X-ray apparatus will clearly show whether there is a pneumothorax.

STAB WOUNDS OF THE ABDOMEN

Whereas twenty years ago any stab wound of the abdomen was regarded as a clear indication for operation, recent studies have shown that a more selective approach yields results at least equally good, if not better.

In some centers, where large numbers of these cases are regularly encountered and where neither a policy of invariable laparotomy nor of careful and repeated clinical examination is feasible, simple exploration of the stab wound under local anesthesia has been advocated and has proved successful. The wound must be widened so that adequate inspection of the peritoneum can be made. If there is no penetration of the peritoneum, the wound is left open and the patient can be discharged. If peritoneal penetration is present, the patient is immediately transferred to the operating theater, where formal laparotomy is carried out.

If a selective policy is chosen, the patient is observed at hourly intervals for signs of shock or peritonitis. If evidence of either of these conditions is found, immediate laparotomy is indicated. The emphasis must be on careful, repeated, and frequent examination of the patient if one is to take this approach. At the Charity Hospital in New Orleans, the policy has been shown to be at least as safe as invariable operation.

22. The postoperative abdomen

Few clinical situations are as diagnostically demanding as evaluation of the abdomen in a patient who has undergone an abdominal operation. The superimposition of the recent abdominal incision and of postoperative narcotics on the preoperative condition compounds the already difficult task of diagnosis of acute abdominal pain. The surgeon must question the postoperative patient and examine the abdomen in every case at least twice daily because it is *changes* in symptoms and findings that are crucial.

The uncomplicated postoperative state

Since we are beginning with an abnormal condition, it would be well to review the usual uncomplicated course of the postoperative abdomen.

Pain. In the early postoperative period, pain is usually steady and fairly severe, with exacerbations produced by movement, deep breathing and coughing that cause the patient to lie quietly in bed. The pain is most pronounced in the first twelve to twenty-four hours and then gradually diminishes during the next several days. As it decreases, the steady component disappears first so that the patient is completely comfortable when still. Pain on motion remains, albeit continually dwindling for about two or three weeks, when only the slightest amount of discomfort remains. The aged patient tends to be much more comfortable than the youthful one, especially if the latter is heavily muscled. I have not found that there are differences between transverse and vertical incisions, although upper abdominal incisions tend to be more uncomfortable than lower abdominal wounds.

"Spasms." On occasion, startlingly intense spasmodic incisional and general abdominal pain occurs in the first thirty-six to forty-eight hours. Unless aware of this syndrome, the physician may conclude that some sort of catastrophe has taken place. The syndrome is so classical, however, that it should be readily recognizable. The "spasms" are precipitated by the slightest motion of the patient or bed, or by a cough, and sometimes they occur without provocation. They last for about thirty seconds and are only slightly alleviated by large doses of narcotics and/or muscle relaxants. During a "spasm" the entire abdomen is tensely rigid. Between "spasms" the patient is as comfortable as the usual postoperative patient. The "spasms" are most intense in the early postoperative hours and gradually diminish, disappearing in thirty-six to forty-eight hours. The syndrome is more often encountered in heavily muscled males after upper abdominal incisions and after mass closures but is not restricted to those circumstances. I regard these pains as an extreme exaggeration of the more common exacerbation of pain produced by sudden motion or cough.

"Gas pains." These are a frequent but not invariable symptom in postoperative patients. They are clearly *colicky* in nature and usually antedate by twelve to twenty-four hours the passage of flatus, an event that often relieves the pain. "Gas pains" can be extremely severe, and I have known surgeon-patients who were convinced that intestinal obstruction had supervened. The abdomen often becomes slightly more protuberant at this time, lending some support to that view. "Gas pains" usually occur, if at all, about

forty-eight hours after such operations as cholecystectomy or vagotomy with drainage, or about four days after colonic resections or aortic procedures. Persistence of the "gas pains" or mild distention strongly suggests the presence of partial intestinal obstruction. I believe that many instances of so-called delayed or persistent ileus are in truth episodes of partial intestinal obstruction.

New and different pains at any time in the postoperative course demand urgent and meticulous attention. Even the mildest of new pains is not to be ignored.

Nausea and vomiting. These are very common in the first twelve to eighteen hours, probably as a result of anesthetic drugs. Vomiting usually consists of frequent small emeses of clear or bile-colored material. This is to be distinguished from the vomiting of *acute gastric dilatation,* which is of the overflow type, with small amounts of often dark or blackish material welling up in the mouth of the patient, who is besieged by hiccups. *In the uncomplicated patient, vomiting should not occur after twenty-four hours; if it does appear, complications should be strongly suspected.*

Bowel habits. Constipation and failure to pass flatus are almost uniformly present, and are usually relieved within two to three days after intra-abdominal operations of mild to moderate extent and within four to five days after more extensive procedures. Passage of large amounts of flatus or a large bowel movement are the best indicators that the postoperative adynamic ileus has resolved. Watery stools as frequently as four or five times in twenty-four hours are not uncommon at the completion of the longer type of ileus, but within the first day or two frequent watery stools may mean compromise of the intestinal circulation such as that which occurs with colonic ischemia after colonic operations or with intestinal obstruction. Persistent obstipation beyond five or six days usually indicates that complete intestinal obstruction is present or that some intra-abdominal catastrophe has occurred, whereas the late onset (one week) of diarrhea should suggest partial intestinal obstruction, especially if there is abdominal distention. *The pas-*

sage of stool and gas is not a guarantee that all is well within the peritoneal cavity!

Inspection and palpation. These are the maneuvers of greatest value in the postoperative patient. Careful inspection and palpation should be carried out twice daily with the patient in the same recumbent position. Failure to do so may result in failure to detect early changes, with catastrophic consequences.

Inspection shows prominent guarding and inhibition of the abdominal contribution to respiration in the first twenty-four hours followed by gradual normalization. After forty-eight hours, much of the restriction of motion has disappeared and the abdomen may become slightly more protuberant than it had been, after which all distention will resolve.

Direct tenderness, percussion tenderness, and true rigidity are invariably present early, being most evident in the region of the incision and gradually decreasing as one moves away from the wound. It is in examining the postoperative patient that the surgeon must be gentlest; elicitation of classical rebound tenderness should never be attempted. Tenderness and rigidity recede gradually with time from the periphery toward the incision. This process of recession of tenderness and rigidity should be assessed twice daily with the utmost gentleness; it serves as a baseline determination for the development of any untoward intra-abdominal event. By this means one soon begins to appreciate the subtlest of changes.

Auscultation. Contrary to popular belief, peristaltic sounds are of little value and are often misleading in assessing the abdomen postoperatively. Few peristaltic sounds are heard in the first twenty-four hours, after which any constellation of auscultatory findings may be detected. The gravest of intra-abdominal events may be missed or neglected "because bowel sounds were normal," a comment heard all too often. The return of peristaltic sounds is taken by many to indicate that oral feedings may be resumed, but here too the unwary physician can err.

Auscultation has been of greater value in detecting pneumonia, pulmonary infarction, or atelectasis. Friction rubs over the site of a hepatic biopsy or metastasis, and even occasionally over an area of peritonitis, make listening worthwhile when unusual pain of some type is a problem.

Fever. It is not within the scope of this book to consider the array of causes of postoperative fever. Suffice it to say that careful studies indicate that fever is invariable after all abdominal operations if fever is defined as an oral temperature higher than 99.6°F. In a completely uncomplicated course the oral temperature rarely exceeds 100°F or 100.2°F on the first postoperative night and gradually decreases thereafter. Should the patient have become hypothermic during the course of an extensive aortic or retroperitoneal procedure, the temperature may reach as high as 102°F orally in the first twenty-four hours. The *trend* of the temperature at the same time each day is extremely important, and an increase of only 0.5°F per day for three or four days often portends the presence of an abscess in the wound or within the abdomen.

The complicated postoperative condition

It cannot be overemphasized that *adverse changes* in the symptoms or findings, no matter how *subtle*, are the hallmarks of a serious complication. One frequently hears that an anastomotic disruption with peritonitis caused the sudden appearance of shock and deterioration of a patient, but careful perusal of the history and the chart usually shows that something has been amiss for several days before.

PERITONITIS

Peritonitis is present *early* in the postoperative period in patients who are operated on for that condition and also in those without preoperative peritonitis in whom the peritoneal cavity may have been contaminated during the operation. The latter is especially likely if the obstructed intestine has been entered or if fecal contamination is great during open intestinal anastomosis. *Peritonitis that occurs from the fourth to the eighth or ninth postoperative day is almost always caused by an anastomotic disruption.*

Early peritonitis. The pain of *early* postoperative peritonitis is rarely distinguishable from incisional pain. It is often mitigated or obscured by narcotics and antibiotics. Intraoperative aspiration and dilution of purulent material may temporarily mask the pain of peritonitis. Surprisingly, therefore, pain is not an informative symptom and a lack of severe pain should not be construed to indicate absence of peritonitis. Nausea and vomiting may continue beyond the first twelve to eighteen hours unless a tube is used to aspirate gastric contents. *Persistent abdominal distention* despite gastric aspiration is the most consistent finding. Because fever is frequently masked by antibiotics, the abdominal distention and failure of resolution of the ileus are sometimes misinterpreted as signs of mechanical obstruction. Operation at this juncture on the basis of such a mistaken impression is not only difficult but often life-threatening. It is not generally recognized that severe heartburn is frequently associated with abdominal distention. Increased intra-abdominal pressure may even overcome the resistance of a normal lower esophageal sphincter with resulting heartburn, a symptom that usually indicates the need for gastric and intestinal decompression rather than antacids, which are so often reflexly prescribed. Distention and hiccups may persist for a week or ten days and sometimes even for two weeks if peritonitis has been severe and generalized. Failure of the distention to resolve at this point usually mean that mechanical obstruction and/or intra-abdominal abscess have occurred. The latter should be considered especially if distention and fever are present.

During the early evolving phases of peritonitis *large volumes of extracellular fluid* are required. Tachycardia, hypotension, and oliguria will ensue unless these large requirements are recognized.

Later peritonitis in a previously well patient. This is almost always the result of anastomotic dehiscence.

Pain is usually present and represents a change in the pattern of discomfort. Any new pain or any pain removed from the incision should be regarded with suspicion. The pain of serious anastomotic disruption is not infrequently interpreted erroneously as "gas pains." The patient usually becomes anorectic and may develop heartburn, nausea, and vomiting. The abdomen usually becomes distended, albeit not greatly so in the early phases. Tenderness and rigidity are usually present but may be so mild as to be misleading. Even more deceptively, passage of flatus and movements often continues. *Fever* is one of the most helpful signs. It may be heralded by a rigor and may reach high levels rapidly in cases of sudden massive disruption or it may gradually increase, reaching a higher peak on each succeeding evening. Requirements for large volumes of extracellular fluid, tachycardia, and hypotension are late events and usually indicate delay in diagnosis. A definite diagnosis can usually be made by means of water-contrast radiopaque media in the case of upper gastrointestinal or colonic anastomosis. These techniques should be used when there is any question of anastomotic failure because they are safe and helpful. The diagnosis must sometimes rest solely on clinical grounds in cases of radiologically inaccessible anastomoses such as those of the biliary tract to defunctioned intestine.

DIFFERENTIAL DIAGNOSIS

1. *"Gas pains."* These are the most common "new" pains that occur in a previously well postoperative patient, and most often they need to be differentiated from anastomotic disruption. They can usually be distinguished by their intermittent and cramping nature and by the lack of fever.
2. *Intestinal obstruction.* Early postoperative obstruction is differentiated from anastomotic disruption in the same way that "gas pains" are.

3. *Wound infection.* This may cause pain and fever with abdominal tenderness, but the pain and tenderness are localized to the incision. The latter may look remarkably uninflamed and normal for several days before the development of overt redness or fluctuance.
4. *Other abdominal conditions.* The postoperative patient seems especially likely to develop *acute cholecystitis* even in the absence of biliary calculi. *Perforated ulcer* or *acute appendicitis* may occur and should always be kept in mind.

ABSCESS

Intra-abdominal abscesses develop after a bout of generalized peritonitis or when extensive contamination of the virgin peritoneal cavity has occurred during the operation, especially if hemostasis has been less precise than it should be. They almost never occur after *clean* intraperitoneal operations unless prolonged drainage has been employed or a foreign body such as a sponge has been left behind. The presence of an abscess usually becomes evident within a week or ten days after the operation, but it is sometimes delayed for weeks or even months by the promiscuous use of antibiotics or steroids.

SIGNS AND SYMPTOMS

1. *Fever.* Fever may be the sole finding. Its natural course should not be obscured by antipyretics or antibiotics unless a specific diagnosis has been established because these agents can delay the diagnosis by months or years. The importance of *repeated examinations* cannot be overemphasized. Fever may continue for days or weeks until the abscess can be detected by physical examination or other means.
2. *Pain and tenderness.* These may be remarkably mild, though they are usually present to some degree. It is surprising how often pain and tenderness are absent in patients with sub-

diaphragmatic abscess. A careful search for pain and tenderness should be made at least twice daily in a patient with fever suspected of harboring an abscess.

3. *Mass.* The development of a mass not previously detectable in a patient at high risk almost invariably confirms the diagnosis. The mass may be present on abdominal examination, but careful search by pelvic and/or rectal examination should also be made.

Other examinations. Radiological examination may show a neighborhood pleural effusion, limitation of diaphragmatic motion, and elevation of the diaphragm in cases of subphrenic abscess. Occasionally, air-fluid levels can also be seen on ordinary X-rays. In recent years, ultrasonography and CT have aided the diagnosis. None of these tests is without false positives and false negatives, but they are often of great help not only in confirming the diagnosis but also in establishing very precisely the location of the lesion.

INTESTINAL OBSTRUCTION

It is my opinion that so-called delayed or late adynamic ileus is in fact often an episode of partial small-bowel obstruction. The patient who becomes distended and uncomfortable "because we fed him too soon" or "because tube decompression was discontinued too soon" very often has low-grade small-bowel obstruction.

Signs and symptoms. These may be not different from those of any patient with intestinal obstruction, except that postoperative obstruction is more *often incomplete*. For that reason, the signs and symptoms may be confined to heartburn and distention. In addition, frequent watery stools (as many as ten to twelve per day) with little or no passage of gas is common and sometimes misleading. *In the presence of fever, partial or complete obstruction in the postoperative patient suggests the presence of an abscess. Radiological examinations* rarely differentiate partial small-bowel obstruction from protracted adynamic ileus since gas is present in the small and large intestines in both instances. Occasionally, however, plain and upright films of the abdomen can demonstrate clearly that small-bowel obstruction exists if the small intestine is greatly dilated at the same time that there is an absence of gas in the colon. Conversely, the demonstration of colonic gas alone in a distended patient with "gas pains" may serve to differentiate these pains from true mechanical obstruction.

23. The acute abdomen in the tropics

Every surgeon intending to practice in the tropics needs to have a good knowledge of tropical medicine and must be on guard when one of the main symptoms is acute abdominal pain, for the common types of acute abdominal crises differ from those seen in temperate climes. In general, the pattern of acute abdomimal disease varies with the climate and sanitary conditions, and with the diet and habits of the inhabitants. Some of the acute inflammatory conditions, for example, acute appendicitis and acute cholecystitis, are less common in the tropics, particularly in the remoter regions and among the native races. On the other hand, there are a number of acute abdominal conditions common in the tropics but seldom seen elsewhere. Special mention must be made of those conditions consequent on *malaria, sickle-cell anemia, amebiasis, infestation with worms,* and *the results of excessive heat.*

MALARIA

Malaria is so common and its symptoms are so protean that it must be considered as a possible cause in any doubtful case in which acute abdominal pain is one of the symptoms. Malarial symptoms have been misdiagnosed as due to one or another of the following: *amebic* or *bacillary dysentery, cholera, intestinal obstruction, appendicitis, cholecystitis, hemorrhagic pancreatitis,* and *liver abscess.* The golden rule, therefore, is to make the necessary microscopic examination of the blood in any doubtful case of acute pain in the abdomen. Nevertheless, the finding of the malarial parasite in the blood does not exclude other actue abdominal conditions, though it makes it incumbent on the observer to make doubly sure that the symptoms are decisive of that condition. Malaria is by far the most common cause of spontaneous rupture of the spleen and predisposes to rupture of that organ.

SICKLE-CELL ANEMIA

Sickle-cell anemia may be accompanied by painful abdominal crises often precipitated by a viral or bacterial infection. The condition occurs only in blacks and is more common in children of that race. The pain is felt in the upper or central abdomen. The onset of the pain may be gradual, vague, and generalized or acute and prostrating, accompanied by abdominal tenderness, nausea, or vomiting. Sometimes there is rigidity of the abdominal wall. Most patients sweat freely and complain of a severe headache, while inquiry may elicit the fact that they have had similar previous attacks and perhaps may have suffered from joint pains of the rheumatic type. Therefore, when dealing with nonwhite patients, suffering from abdominal pain of doubtful origin, it is recommended that the physician examine a blood slide and carry out a test for blood sickling as a routine. Patients with sickle-cell anemia are often icteric and have an increased tendency to form gallstones. Crises with or without hepatic infarcts are often difficult to distinguish from acute cholecystitis.

AMEBIASIS

Amebiasis is met with in most tropical countries and frequently has to be considered in the diagnosis of acute abdominal conditions. It may occur in many forms, so that it is essential for the practitioner to know the common symptoms. Sometimes amebic infection of the large bowel leads to diarrhea or dysentery with colicky pain, but frequently there is no diarrhea and the first intimation of infection may be the onset of hepatitis. Sometimes, even the symptoms of hepatitis may be slight and an abscess of the liver may develop insidiously and lead to general loss of health without any attention being called to the liver, or the silent abscess may give the first intimation of its presence by rupturing into the peritoneal cavity, with resulting acute abdominal symptoms.

Acute hepatitis of amebic origin may develop during the course of acute amebic dysentery or may follow long after the acute dysenteric attack has subsided. The amebae travel to the liver along the radicles of the portal vein and, lodging in the portal capillaries, may cause extensive inflammation.

Symptoms of amebic hepatitis. There are five constant features and many inconstant symptoms to be looked for in amebic hepatitis. The constant features are:

1. Fever.
2. Enlargement of the liver.
3. Pain and tenderness in the hepatic region.
4. Leukocytosis.
5. Rapid subsidence of the symptoms under treatment by emetine or other amebacidal drugs.

The common though inconstant features are:

6. A history of an attack of dysentery.
7. Pain on top of the right shoulder or in the right iliac region.
8. Jaundice.
9. A rigor.
10. Malaise and general lassitude.

The occurrence of fever, with an enlarged painful liver, in a person who is or has been living in a tropical country, should always lead one to suspect amebic hepatitis, whether or not there is a definite history of dysentery. The presence of a leukocytosis of 15,000–20,000 is corroborative evidence. (In very anemic patients a count of much less than that number would be regarded as a leukocytosis.) The diagnosis is always clinched by the rapid, often dramatically rapid, subsidence of symptoms under treatment by amebacidal drugs. The symptoms begin to abate after the giving of two or three doses. If the symptoms do not subside under this treatment, one should again consider malarial hepatitis or even cholecystitis as the possible cause. A chest X-ray usually shows elevation of the right hemidiaphragm. There is a particularly prominent hump of the diaphragm anteriorly in the presence of amebic abscess.

Amebic infection may give rise to painful swellings in the cecal region simulating appendicitis or to a tumor-like mass in the rectum that may give rise to a suspicion of a cancer of the rectum. Similar diagnostic precautions will prevent any unnecessary operation.

INFESTATION WITH WORMS

Infestation with worms is common in many parts of the tropics where sanitary precautions are deficient or absent, particularly in the central zone of Africa, in India, and in southern China. Abdominal symptoms are frequently caused by infestation with round-worms (*Ascaris*), which may multiply in the jejunum until a large mass results and may cause intestinal obstruction. More rarely, one or more worms may penetrate the wall of the intestine and cause local peritonitis or an abscess. *Ascaris* may also obstruct the appendix or the common bile duct, with consequent cholangitis.

Filariasis is prevalent in some parts of India and when it affects the retroperitoneal lymphatics on the right side, it may simulate appendicitis. Fever to 103°–104°F with a rigor usually precedes pain, which usually emanates from the right iliac fossa. Nausea and vomiting usually occur. There is often mild tenderness but no guarding or rigidity. There may or may not be associated epidymoorchitis.

THE EFFECTS OF HEAT

In very hot seasons and in places where a number of patients are being affected by heatstroke, it is well to remember that the initial stage of some cases of heatstroke may exhibit abdominal symptoms. It is seldom, however, that there could be any mistaking of these symptoms for an acute abdominal crisis, for there are always other symptoms—dry skin, hyperpyrexia, mental dullness, etc.—that would guide one correctly.

ABDOMINAL SYMPTOMS DUE TO SALT DEFICIENCY

In tropical climes, as the result of the great heat and consequent excessive sweating, there may result a state of salt deficiency. This may give rise to severe cramps that may affect the muscles of the abdominal wall and cause such great pain, collapse, rigidity, and tenderness that the observer may even diagnose the condition as one due to perforation of a peptic ulcer. A quick test of the serum electrolytes and hematocrit should suffice to make the diagnosis so that prompt treatment can be instituted. Symptoms usually subside quickly upon administration of intravenous sodium chloride solutions.

PERITONITIS

In the tropics the causes of peritonitis are different from those in temperate climes.

The special types of peritonitis to be remembered are pneumococcal peritonitis and peritonitis caused by rupture of an amebic abscess or consequent on the perforation of a typhoidal or dysenteric ulcer. In addition, one must mention peritonitis due to perforation of the bowel by a roundworm. Gonococcal peritonitis is frequently found in women who have not been adequately treated for the infection.

Pneumococcal peritonitis, which in temperate climes is seldom seen except in young girls, in tropical Africa is more common in adult women; for diagnostic purposes in doubtful cases aspiration of the peritoneal cavity through a small, hollow needle is recommended. By staining and examining under the microscope even as small an amount as one drop of the peritoneal exudate, one can differentiate peritonitis due to the pneumococcus from that due to intestinal perforation.

Rupture of an amebic abscess of the liver into the general peritoneal cavity is not common, but in those parts where amebiasis is very common it should be constantly borne in mind.

Rupture of a typhoidal or dysenteric ulcer is not so common now as it was before the recent methods of treatment of the primary complaints, but from time to time in untreated cases it will be met with. It is sufficient to remind the observer of its possibility.

In *perforation of the small bowel by a roundworm*, if the leakage is very slight, the opening may heal spontaneously or a small local abscess may result. If the inflammation extends, general peritonitis may result. The initial pain is usually felt slightly to the left of the umbilicus.

In those parts of the tropics where there are populous cities and a more plentiful food supply, relatively more cases of appendicitis are to be expected. Perforation of a gastric or duodenal ulcer, though not unknown, is uncommon in the tropics.

With very acute *cholera sicca*, the very violent acute abdominal pain may precede the choleraic diarrhea, but the collapse and pain are not accompanied by abdominal rigidity.

PYOMYOSITIS

Pyomyositis or suppurative inflammation of the muscles may affect the muscles of the abdominal wall and cavity and simulate an intra-abdominal crisis. If the parietal muscles are involved there will be fever, local tenderness, and swelling of the abdominal wall, with perhaps the involvement of muscles elsewhere, and absence of evidence of a deep-seated lesion should serve to discriminate. When pyomyositis affects the flat part of the iliacus muscle it may easily simulate an appendicular abscess. The conditions from which pyomyositis needs to be distinguished in various parts of the abdomen have been summarized thus:

Inguinal Region:	Lymphosarcoma
	Lymphogranuloma venereum
	Strangulated hernia
Lower Abdomen:	Appendix abscess
	Psoas (tuberculous) abscess
	Hematoma of the rectus sheath
	Extravasation of urine
Upper Abdomen:	Liver abscess
	Subphrenic abscess
	Splenic conditions
	Perinephric abscess

INTESTINAL OBSTRUCTION IN THE TROPICS

Perhaps the most surprising difference between the acute abdominal surgery of temperate and tropical climes is to be found in the large group of obstructions of the intestine. Strangulated hernia is quite common in both climes, but in almost every other respect the surgeon will find unexpected differences. In some parts of Africa and India intussusception may form 20–30 percent of all cases of intestinal obstruction, while in other parts volvulus of either the small or large bowel may form more than a fifth of all such cases. Cancer is a very rare cause of obstruction in the tropics, but in its place *obstruction of the small bowel by large masses of roundworms* may cause great difficulty in diagnosis and evoke surprise on opening the abdomen. These masses sometimes form a tumor palpable through the abdominal wall.

Intussusception and volvulus appear to be more common among those populations whose diet consists largely of vegetable food with much indigestible residue. In cities where the diet is more varied, neither intussusception nor volvulus is so common. There is great variability in the incidence of the different forms of acute intestinal obstruction in different parts of central Africa, intussusception in adults and volvulus of the small and large intestines being more common, and obstruction due to malignant disease of the colon being much rarer, than in temperate climes.

Similar differences in the incidence of the different forms of acute abdominal crises have been recorded by Professor I. W. J. McAdam from the Mulago Hospital, Kampala, Uganda, where from 1958 to 1960 there were 841 acute nontraumatic abdominal emergencies, of which 794 were due to intestinal obstruction, 14 to perforated peptic ulcer, 21 to appendicitis, 7 to cholecystitis, 1 to renal calculus and 4 to acute pancreatitis. Of the cases of intestinal obstruction 594 were due to strangulated external hernia, 110 to volvulus, 30 to intussusception, 32 to adhesions and bands, 8 to adult pyloric stenosis, and 10 each to congenital abnormalities and malignant disease.

The common form of intussusception seen in adults usually forms a sausage-shaped tumor, does not cause complete obstruction, and does not give rise to blood in the stools. The symptoms of sigmoid volvulus are as given in Chapter 13. The symptoms of volvulus of the small bowel, and of the double volvulus in which the small and large bowels are intertwined, are much more acute, with severe pain, collapse, and rapid onset of gangrene in the closed loop, unless timely operation averts that serious condition.

REGIONAL DIAGNOSIS OF ABDOMINAL CRISES IN THE TROPICS

Acute right hypochondriac pain and tenderness. Acute hepatitis of amebic origin should first be excluded and the possibility of malarial hepatitis considered. Tropical myosis should be excluded by examining muscles other than those of the abdominal wall. If these can be excluded the condition may be due either to a cholecystitis or to a leaking duodenal ulcer.

Pain in the right lower quadrant of the abdomen. Here the probabilities in the tropics are very different from those in temperate climes, for appendicitis is comparatively rare among the native races. Appendicitis is also uncommon in India and is seldom seen in the remoter parts of Africa or China. In large cities there is usually a higher incidence of appendicitis.

In tropical regions, therefore, pain and tenderness in the right lower abdomen are not so likely to be due to appendicitis, and one must consider as more likely causes amebic colitis, leaking liver abscess, intussusception, and pyomyositis of the iliopsoas muscle. The insidious nature of amebic colitis permits a considerable pathological lesion to exist without many symptoms. Ulceration in the cecum may be accompanied by considerable swelling of the gut without dysenteric symptoms. If, in such a case, pain ensues and a definite lump can be felt in the iliac fossa, the conditions for an erroneous diagnosis of appendicitis are present, the more so as there may be slight fever. The examination of the feces for amebae, and the crucial test of treatment, should prevent needless surgical intervention.

Leakage of a liver abscess down the paracolic gutter may cause pain in the right iliac region. Careful examination of the hepatic area should suffice for discrimination from appendicitis. Patients with intussusception give a different history, usually present different symptoms, and often show a movable tumor.

Pain and tenderness in the left upper quadrant. The most common cause of pain and tenderness in the left hypochondrium is injury to or disease of the spleen. Many tropical diseases cause enlargement of the spleen—malaria, kala-azar, sickle-cell trait, and other conditions. Such a diseased spleen can easily be ruptured by slight violence and, apart from disease, the spleen is very liable to injury from any violence applied to the lower left chest or the left upper abdomen. The hemorrhage may be only subcapsular at first, but there is always danger of a later hemorrhage into the general peritoneal cavity.

With a ruptured spleen, in addition to any local signs of injury, there may be a complaint of pain felt on top of the left shoulder (phrenic pain) and the pulse rate may gradually rise. Aspiration of the abdominal cavity may reveal the presence of blood, and the hematocrit reading may indicate hemorrhage. The local pain may be increased when the patient takes a deep breath. Conditions that may need to be differentiated from a ruptured spleen are left-sided pleurisy and a pyomyositis of the muscles of the back and flank.

24. Diseases that may simulate the acute abdomen

There are a number of diseases that either do not need or positively contraindicate operative interference, which may yet cause symptoms very suggestive of conditions for which operation is the best procedure. In some cases the symptoms arise from disease within the abdomen; in other instances the pain is referred to the abdomen from another part of the body (e.g., thorax or spine).

GENERAL DISEASE

It is not uncommon for abdominal pain and vomiting to occur at the onset of some of the specific fevers or of influenza, especially in children, but in such cases the general symptoms outweigh the local manifestations. Fever will be present at the start and the general malaise is greater, while locally there will not be found the tenderness of rigidity that suggest intra-abdominal inflammation, and it is by this discrepancy that the observer will be guided. Occasionally the abdominal symptoms may precede the general manifestations of influenza; the pain may be of a severe colicky type, and if accompanied by vomiting may give rise to a suspicion of acute intestinal obstruction. The vomiting never tends to become feculent and soon the general symptoms of influenza appear. During an epidemic of influenza one is more likely to mistake an acute abdominal condition for an attack of influenza with gastric symptoms—a serious mistake.

BLOOD DISEASES

Acute porphyria. This may be accompanied by attacks of acute abdominal pain that have, on occasion, been mistaken for appendicitis or intestinal obstruction. The condition should particularly be borne in mind when the patient has been taking drugs of the sulfonamide or barbiturate groups (see Chapter 7).

I have known *leukemia* to present as an acute abdominal emergency. The patient gave a history of prolonged indigestion, had suffered recently from irregular fever, and with an anemic appearance presented also great tenderness, rigidity, and dullness in the left hypochondrium, so that the simulation of subphrenic abscess due to a leaking ulcer was rather close. A leukocyte count, however, showed 240,000 white cells to the cubic millimeter and the true condition was recognized. The acute local reaction was due to tension and threatening rupture of the spleen, for spontaneous rupture took place shortly afterward.

Attacks of severe abdominal pain accompanied by vomiting (and sometimes diarrhea) occur in some cases of *pernicious anemia* but seldom give rise to serious difficulty in diagnosis. We have elsewhere mentioned the acute abdominal pains that may occur during the course of sickle-cell anemia in blacks.

HEMOLYTIC CRISES

It is well known that this condition is sometimes accompanied by abdominal crises in which acute abdominal pain, nausea or vomiting, fever, and tenderness on palpating the abdomen are noteworthy features. The attacks coincide with an increase in the depth of the jaundice and the intensity of the anemia. The cause of the abdominal pain during a hemolytic crisis is unknown. Since many of these patients have pigment gallstones, considerable difficulty may arise in diagnosis. The lack of typical features of biliary colic or cholecystitis together with evidence of major hemolysis and anemia usually serves to distinguish hemolytic crises.

THORACIC DISEASES

Pleurisy or pleuropneumonia. Either of these conditions may cause abdominal pain and rigidity and may be accompanied by vomiting. Whether the pleurisy results from pneumonia or a pulmonary infarct, abdominal pain is often present. In some cases it is quite easy to distinguish these from abdominal conditions by the signs present in the chest, but in children in whom the thoracic signs are often late in appearing, and in some cases of diaphragmatic pleurisy in which few signs may be found, diagnosis is extremely difficult. A tabulation of the main differential points is given on page 283. An additional test is illustrated in Figure 35. If the pain is unilateral and of abdominal origin, pressure from the opposite side of the abdomen toward the affected side will cause pain, while if the pain is referred from the thorax, no pain is caused by such pressure.

Acute cardiac disease. This frequently causes symptoms referable to the abdomen. Epigastric pain and tenderness are common when the liver is congested and swollen from back pressure, but nausea and vomiting are unusual despite intense pain. Vomiting is not an infrequent symptom in myocardial infarction, and severe collapse may usher in an attack of pericarditis or accompany acute cardiac failure. I have on occasion been called to cases of endocarditis, pericarditis, and coronary thrombosis that were thought to be cases of abdominal disease. Needless to say, in any case of doubt, the circulatory system must be very carefully examined. In some instances doubt remains even after one has measured the cardiac and hepatic dullness, listened to the cardiac sounds, noted carefully the character and rate of the pulse, measured the blood pressure, and observed if there is any venous pulsation. Sometimes the electrocardiogram will be of invaluable assistance in distinguishing between myocardial infarction and biliary colic. However, it may be of little help in deciding between biliary colic and angina pectoris (see Chapter 10).

One case in which I have known doubt was that of a man of sixty to whom I was summoned in the middle of the night for a supposed

COMPARATIVE TABLE OF SYMPTOMS IN ACUTE ABDOMINAL
AND ACUTE PLEURAL OR PNEUMONIC LESIONS

Abdominal	Pleural or pneumonic

Previous history
  Indigestion
  Colicky pains
  Constipation
  Diarrhea
	Common cold or "chill"
	Exposure to infection or previous oper-
	  ation

Onset
  Acute without fever (except pyelitis)
  Rigor unusual (except pyelitis)
  Vomiting usual
  Pain often shifts downward
	Acute with fever at start
	Rigor common
	Vomiting less common
	Pain thoracic as well as abdominal

Examination

Abdominal	Pleural or pneumonic

Appearance
  Varies from normal to "abdominal"
    facies
	Cheeks flushed
	Alae nasi working
	Sometimes herpes on lips

Skin Cold or clammy or normal	Skin may be hot and dry
Pulse and respiration No sure guide	P:R ratio lessened
Abdominal wall May be rigid	Less commonly rigid
Phrenic shoulder pain Common, but seldom below clavicle	Common, especially below clavicle
Psoas test Often positive	Always negative
Obturator test Sometimes (rarely) positive	Always negative
Testicular pain Sometimes present	Never present

Rectal examination
  May elicit tenderness or demonstrate
    lump
	Negative

Examination of chest
  Frequently slight rubs in upper ab-
    dominal lesions
	May be a rub or dullness or bronchial
	  breathing, but sometimes nothing
	  definite at onset of symptoms

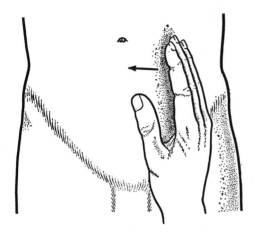

Fig. 35. The method of differentiating between right-sided abdominal pain of thoracic and abdominal origin. (Pressure from the left side causes pain if abdominal in origin.)

perforation of a gastric ulcer. The patient had been under treatment for twelve months for gastric ulcer, which caused frequent attacks of indigestion. Sudden abdominal pain and collapse had supervened several hours before I was summoned. I found the patient fully conscious and very talkative, but extremely distressed and short of breath. No pulse could be felt at either wrist, nor could the brachial arteries be felt to pulsate. The heart was beating 160 to the minute, and the superficial veins of the neck were pulsating visibly. The liver was tender and enlarged down to the umbilicus. Pain was felt down the left arm to the elbow. I diagnosed cardiac failure and angina pectoris and refused to operate. The patient died five hours later, and a postmortem examination of the abdominal cavity showed no gastric ulcer and no peritonitis. The thorax was not examined, but it was clear that the frequent attacks of indigestion had been slight attacks of angina pectoris. The diagnosis of cardiac failure would not have given rise to so much doubt if the gastric lesion had not been so confidently diagnosed previously.

Ordinary cardiac failure does not cause true rigidity of the abdominal wall, though pressure on a tender, swollen liver may elicit muscular resistance. With acute pericarditis, however, there may be true rigidity of the abdominal wall, but this would be unaccompanied by other signs of abdominal disease. With pericarditis also there may be phrenic referred pain felt under the left clavicle. The significant pain down the left arm sometimes felt in cardiac disease may be helpful in diagnosis.

DISEASE OF THE SPINE OR SPINAL CORD

Osteomyelitis. Acute osteomyelitis of the dorsal or lumbar vertebrae may cause abdominal pain and rigidity, but there will be great tenderness on pressure over the affected part of the spine that will draw attention to the origin of the pain.

In children acute abdominal (epigastric or umbilical) pain may be consequent on *Pott's disease of the spine.* The absence of abdominal signs would naturally cause examination of the spine and detection of the spinal disease.

Tabes dorsalis. This condition, now rarely seen, causes severe abdominal pain in the form of *gastric crises.* The crises, though more common in adults, may also occur in children who are the subject of juvenile tabes. The pains may be very severe, and uncontrollable vomiting may occur. The important point to remember is that the abdominal wall is not rigid in the intervals of the pain of a gastric crisis.

When there is the slightest doubt about the diagnosis, it should be made a rule to test the pupillary reactions and the knee jerks, so that the mistake will not be made of advising that an abdomen be opened when the pains are caused by tabes dorsalis. It must be recollected, however, that an acute abdomen may occur in a tabetic subject, and one should not hesitate to advise operation if the local signs are definite.

Herpes zoster. Herpes zoster affecting any of the lower dorsal spinal roots may cause pain referred to the anterior or lateral parts of the abdominal wall. When the pain precedes the vesicular eruption it may give rise to the suspicion of intra-abdominal trouble in need of surgery. The pain is sharply delimited by the segmental distribution of the affected nerves and, unless there are other signs, the appearance of the vesicular rash will explain the nature of the pain.

RENAL DISEASE

Serious disease of the kidneys may cause uremia, which may be accompanied by vomiting and abdominal distention. Thus intestinal obstruction may be closely simulated. This may occur in acute nephritis, chronic nephritis, bilateral cystic disease of the kidneys, or pyonephrosis.

If in every case of intestinal obstruction one remembers the possibility of uremia, there should be no great difficulty in diagnosis by considering the differential points set out below:

Intestinal obstruction	Uremia simulating obstruction
Vomiting tends to become feculent	Vomiting not feculent
If obstruction low down, absolute constipation of flatus and feces	Bowels may return flatus after enema
May be history of subacute attack of obstruction	May be history of some surgical or medical disease of kidneys
No renal tumors	In cystic disease bilateral renal tumors found
Blood pressure may be normal	Blood pressure likely to be much raised

It is very seldom that the two conditions are confused when once the possibility of uremia is considered.

PERIARTERITIS NODOSA

Abdominal pain giving rise to a suspicion of appendicitis, cholecystitis, peptic ulcer, or pancreatitis can occur in patients suffering

mucosal pigmentation as well as vitiligo may be present. In addition, the serum sodium is low, the serum potassium high, and the hydroxycorticoid levels low. In the more common variety caused by withdrawal of exogenous steroids, these test results are often normal and therefore misleading. In any case, this latter group of patients must be given steroids, which eliminate the abdominal pain within a few hours.

25. Acute abdominal pain in the immunocompromised patient

The recent epidemic of acquired immunodeficiency syndrome (AIDS) and the increasingly widespread use of immunosuppressive agents to prevent rejection after transplantation and to treat other diseases have led to the recognition that these groups of patients are peculiarly susceptible to certain diseases not frequently encountered in the general population. While the history of an organ transplant or chemotherapy for malignancy is not easily overlooked, the same cannot be said for the possibility that a patient might have acquired AIDS. Those at high risk for AIDS include homosexual or bisexual men, prostitutes, intravenous drug users, recipients of blood products including hemophiliacs, and Zairians both in Europe and in Africa. The presence of unexplained lymphadenopathy, especially if found in two or more extrainguinal sites, or a history of pneumonia caused by *Pneumocystis* should heighten the suspicion of the clinician caring for the patient. Anorectal pathology including anorectal abscess or fistula, or gonorrheal proctitis, is common in individuals with AIDS. It is not within the purview of this chapter to discuss separately each of the acute abdominal emergencies that may be encountered in the immunocompromised patient since these conditions have been described in previous chapters. Rather, the reader's attention is drawn to the wide variety of conditions that are encountered in this population of patients.

Suppression of signs and symptoms

As in individuals on steroids (see Chapter 23), signs and symptoms are often minimal, even in the presence of severe peritonitis. The slightest degree of pain, tenderness, or abdominal distention must be carefully evaluated and regarded very seriously. Elevations in body temperature tend also to be of lesser degree in immunocompromised than in non-immunosuppressed individuals. In many instances, it will only be the slight but definite worsening of subtle signs and symptoms, adduced over a period of hours of evaluation and reevaluation, that will direct the surgeon to an urgently needed laparotomy.

Perforations of the gastrointestinal tract

Perforations of the gastrointestinal tract may develop as a consequence of either unusual infections or tumors, or in some instances without known cause.

Infection with the cytomegalovirus is one of the more common causes of perforation of the colon in these patients, sometimes in association with the clinical picture of toxic megacolon. *Cryptosporidium* may produce similar perforations. Caseating granulomas caused by *Mycobacterium tuberculosae* or the more unusual opportunistic *M. aviare intracellulare* (MAI) have been reported to perforate the distal small intestine. In fact, *M. aviare* infections are more common in AIDS patients than are the usual *M. tuberculosae* infections, and such MAI infections are often widely disseminated. Obviously, perforations in association with the more common conditions such as acute appendicitis and diverticulitis are also encountered.

Lymphomas and Kaposi's sarcoma are the most common tumors in AIDS patients and are sometimes the site of intestinal perforation. The pigmented cutaneous lesions of widespread Kaposi's sarcoma are commonly found around the face, head, neck, and even the oral cavity. Their presence suggests the presence of visceral Kaposi's sarcoma.

Other conditions

Acute cholecystitis is common and is often of the acalculous type in immunocompromised patients. Its diagnosis must be based largely on the history and the physical examination since ultrasonography and cholecystography may be misleading in the absence of stones. The gallbladder wall and/or bile may be infected with cytomegalovirus, *Candida albicans*, *Salmonella*, or *Cryptosporidium*. Varying degrees of jaundice are not uncommonly caused by multiple strictures of the common bile duct and intrahepatic biliary radicles that are the result of infection by *Cryptosporidium*. If the presence of such strictures is recognized, they should *not* be attacked surgically.

Intestinal obstruction may be caused by direct occlusion of the intestinal lumen or by intussusception of a Kaposi's sarcoma or lymphoma, or in some instances by a granuloma.

Splenic abscesses due to *Candida* or *Salmonella* are not uncommon, and they may rupture and produce the picture of localized or generalized peritonitis. This complication usually requires splenectomy to prevent further seeding of the peritoneal cavity. Primary peritonitis of unknown cause has been described in AIDS patients, as has peritonitis secondary to rupture of inflamed mesenteric nodes caused by *toxoplasma*.

Pneumonia caused by *Pneumocystis carinii* is extremely common in AIDS patients. It is important to recognize this condition and to avoid unnecessary operation for the treatment of pain referred from the pleura, particularly because the infection responds reasonably well to appropriate antibiotics. However, acute abdominal pain of a new type arising in a patient with this type of pneumonia raises the possibility of any of the conditions described above.

Index